Student Solutions Manual for Burzynski's
Applied Calculus for Business, Life, and Social Sciences

Prepared by

Ross Rueger
Department of Mathematics
College of the Sequoias
Visalia, California

Student Solutions Manual for Burzynski's
Applied Calculus

Ross Rueger

Publisher: XYZ Textbooks

Sales: Amy Jacobs, Richard Jones, Bruce Spears, Rachael Hillman

Cover Design: Rachel Hintz

ISBN-13: 978-1-63098-049-8 / ISBN-10: 1-63098-049-8

For product information and technology assistance, contact us at
XYZ Textbooks, 1-877-745-3499

For permission to use material from this text or product,
e-mail: info@mathtv.com

XYZ Textbooks
1339 Marsh Street
San Luis Obispo, CA 93401
USA

For your course and learning solutions, visit www.xyztextbooks.com

Printed in the United States of America

Contents

Preface

This *Student Solutions Manual* contains complete solutions to all odd-numbered exercises, and complete solutions to all chapter test exercises, of *Applied Calculus for Business, Life, and Social Sciences* by Denny Burzynski. I have attempted to format solutions for readability and accuracy, and apologize to you for any errors that you may encounter. If you have any comments, suggestions, error corrections, or alternative solutions please feel free to send me an email (address below).

Please use this manual with some degree of caution. Be sure that you have attempted a solution, and re-attempted it, before you look it up in this manual. Mathematics can only be learned by *doing*, and not by observing! As you use this manual, do not just read the solution but work it along with the manual, using my solution to check your work. If you use this manual in that fashion then it should be helpful to you in your studying.

I would like to thank Jennifer Thomas and Amy Jacobs at XYZ Textbooks for their help with this project and for getting back to me with corrections so quickly. Producing a manual such as this is a team effort, and this is an excellent team to work with.

I wish to express my appreciation to Denny Burzynski and Pat McKeague for asking me to be involved with this textbook. I think you will find this text very readable with excellent examples. I especially appreciate thier efforts through XYZ Textbooks to make textbooks more affordable for our students to purchase.

Good luck!

Ross Rueger
College of the Sequoias
rossrueger@gmail.com

August, 2015

Chapter 1
Functions, Limits, and Rates of Change

1.1 Introduction to Functions and Relations

1. The domain is {1,3,5,7} and the range is {2,4,6,8}. This is a function.
3. The domain is {0,1,2,3} and the range is {4,5,6}. This is a function.
5. The domain is {a,b,c,d} and the range is {3,4,5}. This is a function.
7. The domain is {a} and the range is {1,2,3,4}. This is not a function.
9. Yes, since it passes the vertical line test.
11. Yes, since it passes the vertical line test.
13. The domain is $\left\{ x \mid -5 \le x \le 5 \right\}$ and the range is $\left\{ y \mid 0 \le y \le 5 \right\}$.
15. The domain is $\left\{ x \mid -5 \le x \le 3 \right\}$ and the range is $\left\{ y \mid y = 3 \right\}$.
17. Evaluating the function: $f(2) = 2(2) - 5 = 4 - 5 = -1$
19. Evaluating the function: $f(-3) = 2(-3) - 5 = -6 - 5 = -11$
21. Evaluating the function: $g(-1) = (-1)^2 + 3(-1) + 4 = 1 - 3 + 4 = 2$
23. Evaluating the function: $g(-3) = (-3)^2 + 3(-3) + 4 = 9 - 9 + 4 = 4$
25. Evaluating the function: $g(a) = a^2 + 3a + 4$
27. Evaluating the function: $f(a+6) = 2(a+6) - 5 = 2a + 12 - 5 = 2a + 7$
29. Evaluating the function: $f\left(\dfrac{1}{3}\right) = \dfrac{1}{\dfrac{1}{3} + 3} = \dfrac{1}{\dfrac{10}{3}} = \dfrac{3}{10}$
31. Evaluating the function: $f\left(-\dfrac{1}{2}\right) = \dfrac{1}{-\dfrac{1}{2} + 3} = \dfrac{1}{\dfrac{5}{2}} = \dfrac{2}{5}$
33. Evaluating the function: $f(-3) = \dfrac{1}{-3 + 3} = \dfrac{1}{0}$, which is undefined
35. a. Evaluating the function: $f(a) - 3 = a^2 - 4 - 3 = a^2 - 7$
    b. Evaluating the function: $f(a-3) = (a-3)^2 - 4 = a^2 - 6a + 9 - 4 = a^2 - 6a + 5$
    c. Evaluating the function: $f(x) + 2 = x^2 - 4 + 2 = x^2 - 2$
    d. Evaluating the function: $f(x+2) = (x+2)^2 - 4 = x^2 + 4x + 4 - 4 = x^2 + 4x$
    e. Evaluating the function: $f(a+b) = (a+b)^2 - 4 = a^2 + 2ab + b^2 - 4$
    f. Evaluating the function: $f(x+h) = (x+h)^2 - 4 = x^2 + 2xh + h^2 - 4$

37. The domain is all real numbers and the range is $\{y \mid y \geq -1\}$. This is a function.

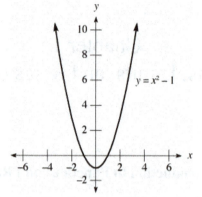

39. The domain is $\{x \mid x \geq -1\}$ and the range is all real numbers. This is not a function.

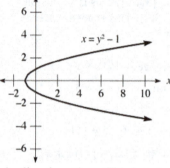

41. The domain is $\{x \mid x \geq 0\}$ and the range is all real numbers. This is not a function.

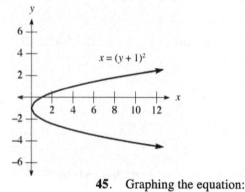

43. Graphing the equation:

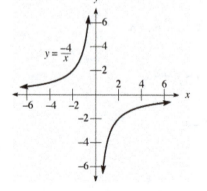

45. Graphing the equation:

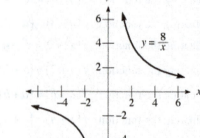

47. Graphing the equation:

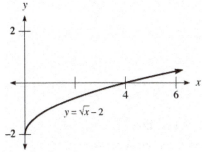

$y = \sqrt{x} - 2$

49. Graphing the equation:

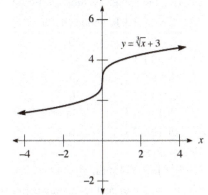

$y = \sqrt[3]{x} + 3$

51. Graphing the function:

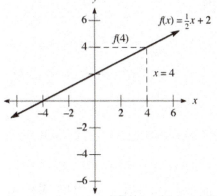

$f(x) = \frac{1}{2}x + 2$

$f(4)$

$x = 4$

53. Graphing the function:

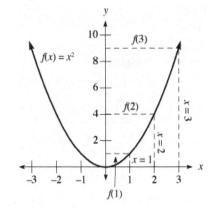

$f(x) = x^2$

$f(3)$

$f(2)$

$f(1)$

$x = 3$

$x = 2$

$x = 1$

55. **a.** The equation is $y = 9.5x$ for $10 \le x \le 40$.

b. Completing the table:

Hours Worked	Function Rule	Gross Pay ($)
x	$y = 9.5x$	y
10	$y = 9.5(10) = 95$	95.00
20	$y = 9.5(20) = 190$	190.00
30	$y = 9.5(30) = 285$	285.00
40	$y = 9.5(40) = 380$	380.00

c. Constructing a line graph:

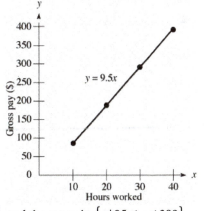

$y = 9.5x$

Gross pay ($)

Hours worked

d. The domain is $\{x \mid 10 \le x \le 40\}$ and the range is $\{y \mid 95 \le y \le 380\}$.

e. The minimum is $95 and the maximum is $380.

57. The domain is $\{x \mid 450 \leq x \leq 2,600\}$ and the range is $\{y \mid 10 \leq y \leq 40\}$.

59. a. Finding the values by substituting x-values:

Year	x	Revenue (millions)
2007	0	0
2008	1	$16
2009	2	$6
2010	3	0
2011	4	$4
2012	5	0

b. Evaluating:

$R(6) = -\$54$ million $R(16) = -\$29,744$ million

No, it can not expect its revenues to rise after 2012, since the revenues are negative.

61. a. True   **b.** False
c. True   **d.** False
e. True

63. Completing the table:

Weeks x	Weight(lb) $W(x)$
0	200
1	194.3
4	184
12	173.3
24	168

65. a. The domain of this function is $\{x \mid x \geq 0\}$.   **b.** Evaluating: $\bar{C}(280) = \dfrac{6(280) + 8,000}{2(280) + 300} \approx \$11.26$

67. $f(x) = 0$ if $x = 0, \pm 2$ **69.** $f(x) = x$ if $x = -1, 2$

71. Sketching the graphs:

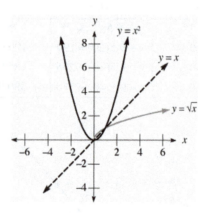

73. a. The domain is $\{x \mid x \geq 0\}$.

b. The domain is $\{x \mid x \geq 2\}$.

c. The domain is $\{x \mid x \geq -2\}$.

75. Sketching the graphs:

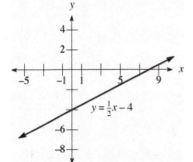

Building a table:

x	y
-5	-6.5
-4	-6
-3	-5.5
-2	-5
-1	-4.5
0	-4
1	-3.5

77. Sketching the graphs:

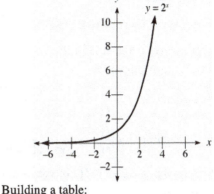

Building a table:

x	y
-2	0.25
-1	0.5
0	1
1	2
2	4
3	8
4	16

79. Multiplying: $0.6(M-70)=0.6M-42$

81. Multiplying: $(4x-3)(4x^2-7x+3)=16x^3-28x^2+12x-12x^2+21x-9=16x^3-40x^2+33x-9$

83. Simplifying: $(4x-3)+(4x^2-7x+3)=4x-3+4x^2-7x+3=4x^2-3x$

85. Simplifying: $(4x^2+3x+2)+(2x^2-5x-6)=4x^2+3x+2+2x^2-5x-6=6x^2-2x-4$

87. Simplifying: $4(-1)^2-7(-1)=4(1)-7(-1)=4+7=11$

1.2 Algebra and Composition with Functions

1. Writing the formula: $f+g=f(x)+g(x)=(4x-3)+(2x+5)=6x+2$

3. Writing the formula: $g-f=g(x)-f(x)=(2x+5)-(4x-3)=-2x+8$

5. Writing the formula: $fg=f(x)\cdot g(x)=(4x-3)(2x+5)=8x^2+14x-15$

7. Writing the formula: $g/f=\dfrac{g(x)}{f(x)}=\dfrac{2x+5}{4x-3}$

9. Writing the formula: $g+f=g(x)+f(x)=(x-2)+(3x-5)=4x-7$

11. Writing the formula: $g+h=g(x)+h(x)=(x-2)+(3x^2-11x+10)=3x^2-10x+8$

13. Writing the formula: $g-f=g(x)-f(x)=(x-2)-(3x-5)=-2x+3$

15. Writing the formula: $fg=f(x)\cdot g(x)=(3x-5)(x-2)=3x^2-11x+10$

17. Writing the formula:
$$fh = f(x) \cdot h(x)$$
$$= (3x-5)(3x^2 - 11x + 10)$$
$$= 9x^3 - 33x^2 + 30x - 15x^2 + 55x - 50$$
$$= 9x^3 - 48x^2 + 85x - 50$$

19. Writing the formula: $h/f = \dfrac{h(x)}{f(x)} = \dfrac{3x^2 - 11x + 10}{3x - 5} = \dfrac{(3x-5)(x-2)}{3x-5} = x - 2$

21. Writing the formula: $f/h = \dfrac{f(x)}{h(x)} = \dfrac{3x-5}{3x^2 - 11x + 10} = \dfrac{3x-5}{(3x-5)(x-2)} = \dfrac{1}{x-2}$

23. Writing the formula: $f + g + h = f(x) + g(x) + h(x) = (3x-5) + (x-2) + (3x^2 - 11x + 10) = 3x^2 - 7x + 3$

25. Writing the formula:
$$h + fg = h(x) + f(x)g(x) = (3x^2 - 11x + 10) + (3x-5)(x-2) = 3x^2 - 11x + 10 + 3x^2 - 11x + 10 = 6x^2 - 22x + 20$$

27. Evaluating: $(f+g)(2) = f(2) + g(2) = (2 \cdot 2 + 1) + (4 \cdot 2 + 2) = 5 + 10 = 15$

29. Evaluating: $(fg)(3) = f(3) \cdot g(3) = (2 \cdot 3 + 1)(4 \cdot 3 + 2) = 7 \cdot 14 = 98$

31. Evaluating: $(h/g)(1) = \dfrac{h(1)}{g(1)} = \dfrac{4(1)^2 + 4(1) + 1}{4(1) + 2} = \dfrac{9}{6} = \dfrac{3}{2}$

33. Evaluating: $(fh)(0) = f(0) \cdot h(0) = (2(0) + 1)(4(0)^2 + 4(0) + 1) = (1)(1) = 1$

35. Evaluating: $(f + g + h)(2) = f(2) + g(2) + h(2) = (2(2) + 1) + (4(2) + 2) + (4(2)^2 + 4(2) + 1) = 5 + 10 + 25 = 40$

37. Evaluating: $(h + fg)(3) = h(3) + f(3) \cdot g(3) = (4(3)^2 + 4(3) + 1) + (2(3) + 1) \cdot (4(3) + 2) = 49 + 7 \cdot 14 = 49 + 98 = 147$

39. **a.** Evaluating: $(f \circ g)(5) = f(g(5)) = f(5 + 4) = f(9) = 9^2 = 81$

　　b. Evaluating: $(g \circ f)(5) = g(f(5)) = g(5^2) = g(25) = 25 + 4 = 29$

　　c. Evaluating: $(f \circ g)(x) = f(g(x)) = f(x + 4) = (x + 4)^2$

　　d. Evaluating: $(g \circ f)(x) = g(f(x)) = g(x^2) = x^2 + 4$

41. **a.** Evaluating: $(f \circ g)(0) = f(g(0)) = f(4 \cdot 0 - 1) = f(-1) = (-1)^2 + 3(-1) = 1 - 3 = -2$

　　b. Evaluating: $(g \circ f)(0) = g(f(0)) = g(0^2 + 3 \cdot 0) = g(0) = 4(0) - 1 = -1$

　　c. Evaluating: $(f \circ g)(x) = f(g(x)) = f(4x - 1) = (4x - 1)^2 + 3(4x - 1) = 16x^2 - 8x + 1 + 12x - 3 = 16x^2 + 4x - 2$

　　d. Evaluating: $(g \circ f)(x) = g(f(x)) = g(x^2 + 3x) = 4(x^2 + 3x) - 1 = 4x^2 + 12x - 1$

43. Evaluating each composition:
$$(f \circ g)(x) = f(g(x)) = f\left(\frac{x+4}{5}\right) = 5\left(\frac{x+4}{5}\right) - 4 = x + 4 - 4 = x$$
$$(g \circ f)(x) = g(f(x)) = g(5x - 4) = \frac{5x - 4 + 4}{5} = \frac{5x}{5} = x$$
Thus $(f \circ g)(x) = (g \circ f)(x) = x$.

45. **a.** The function is $M(x) = 220 - x$.

　　b. Evaluating: $M(24) = 220 - 24 = 196$ beats per minute

　　c. The training heart rate function is: $T(M) = 62 + 0.6(M - 62) = 0.6M + 24.8$

　　　　Finding the composition: $T(M(x)) = T(220 - x) = 0.6(220 - x) + 24.8 = 156.8 - 0.6x$

　　　　Evaluating: $T(M(24)) = 156.8 - 0.6(24) \approx 142$ beats per minute

d. Evaluating: $T(M(36)) = 156.8 - 0.6(36) \approx 135$ beats per minute

e. Evaluating: $T(M(48)) = 156.8 - 0.6(48) \approx 128$ beats per minute

47. Solving the equation:
$$650 = 11.5x - 0.05x^2$$
$$0.05x^2 - 11.5x + 650 = 0$$
$$0.05(x^2 - 230x + 13,000) = 0$$
$$0.05(x - 100)(x - 130) = 0$$
$$x = 100, 130$$

They must sell either 100 or 130 DVDs to receive $650 in revenue.

49. Evaluating the functions:
$$f[g(16)] = f(\sqrt{16}) = f(4) = 5 \cdot 4^2 - 4 = 76$$
$$g[f(16)] = g(5 \cdot 16^2 - 16) = g(1264) = \sqrt{1264} = 4\sqrt{79} \approx 35.6$$

51. Evaluating the functions:
$$f[g(2)] = f(4) = \frac{4-4}{4+1} = 0 \qquad g[f(2)] = g\left(\frac{2-4}{2+1}\right) = g\left(-\frac{2}{3}\right) = \left(-\frac{2}{3}\right)^2 = \frac{4}{9}$$

53. **a.** Finding the profit function:
$$P(x) = R(x) - C(x) = (-1.1x^2 + 41.5x) - (0.02x^3 - 0.5x^2 + 10x + 50) = -0.02x^3 - 0.6x^2 + 31.5x - 50$$

b. Evaluating when $x = 19$: $P(19) = -0.02(19)^3 - 0.6(19)^2 + 31.5(19) - 50 = \$194.72$

c. Evaluating when $x = 20$: $P(20) = -0.02(20)^3 - 0.6(20)^2 + 31.5(20) - 50 = \$180$

d. Evaluating when $x = 21$: $P(21) = -0.02(21)^3 - 0.6(21)^2 + 31.5(21) - 50 = \$161.68$

e. The profit is decreasing as the number of units produced and sold increases from 19 through 21.

55. **a.** The function is: $F(x) = x - 50$

b. The function is: $D(x) = x - 0.20x = 0.8x$

c. The function is: $D(F(x)) = D(x - 50) = 0.8(x - 50) = 0.8x - 40$

d. Evaluating when $x = 500$: $D(F(500)) = 0.8(500) - 40 = \$360$

57. Computing the value: $C[p(15)] = C(1.8 + 0.04 \cdot 15^2) = C(10.8) = \sqrt{0.18 \cdot 10.8^2 + 9.6} = \sqrt{30.5952} \approx 5.53$

In 15 years, the daily level of carbon monoxide is approximately 5.53 parts per million.

59. Writing as a fraction: $-0.06 = -\dfrac{6}{100}$

61. Substituting $x = 2$: $y = 2(2) - 3 = 4 - 3 = 1$

63. Simplifying: $\dfrac{1-(-3)}{-5-(-2)} = \dfrac{4}{-3} = -\dfrac{4}{3}$

65. Simplifying: $\dfrac{-1-4}{3-3} = \dfrac{-5}{0}$, which is undefined

67. Solving for y:
$$\frac{y-b}{x-0} = m$$
$$y - b = mx$$
$$y = mx + b$$

69. Solving for y:
$$y - 3 = -2(x + 4)$$
$$y - 3 = -2x - 8$$
$$y = -2x - 5$$

71. Substituting $x = 0$: $y = -\dfrac{4}{3}(0) + 5 = 0 + 5 = 5$

1.3 Slopes, Rates of Change, and Linear Functions

1. Finding the slope: $m = \dfrac{4-1}{4-2} = \dfrac{3}{2}$

3. Finding the slope: $m = \dfrac{2-4}{5-1} = \dfrac{-2}{4} = -\dfrac{1}{2}$

5. Finding the slope: $m = \dfrac{2-(-3)}{4-1} = \dfrac{2+3}{3} = \dfrac{5}{3}$

7. Finding the slope: $m = \dfrac{-9-(-4)}{5-2} = \dfrac{-9+4}{3} = -\dfrac{5}{3}$

9. Finding the slope: $m = \dfrac{-1-5}{1-(-3)} = \dfrac{-6}{1+3} = \dfrac{-6}{4} = -\dfrac{3}{2}$

11. Finding the slope: $m = \dfrac{6-6}{2-(-4)} = \dfrac{0}{6} = 0$

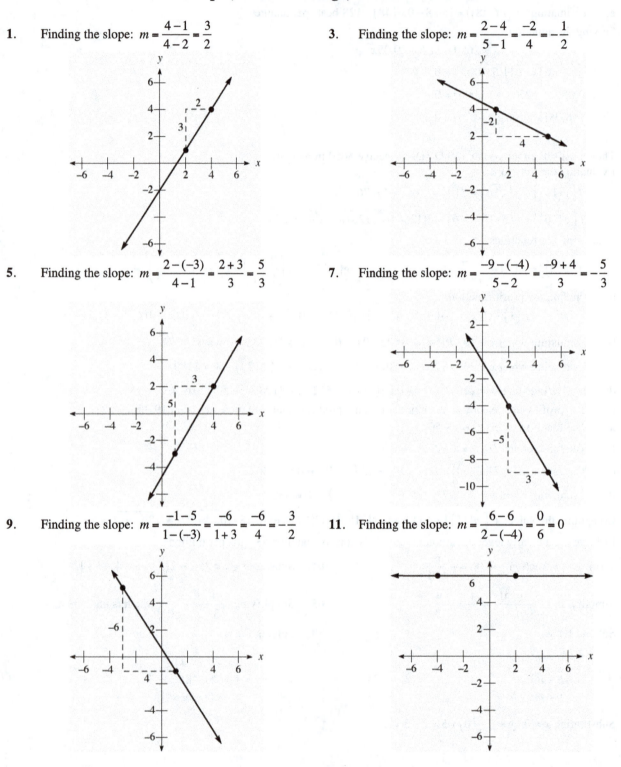

13. Finding the slope: $m = \dfrac{5-(-3)}{a-a} = \dfrac{5+3}{0}$, which is undefined

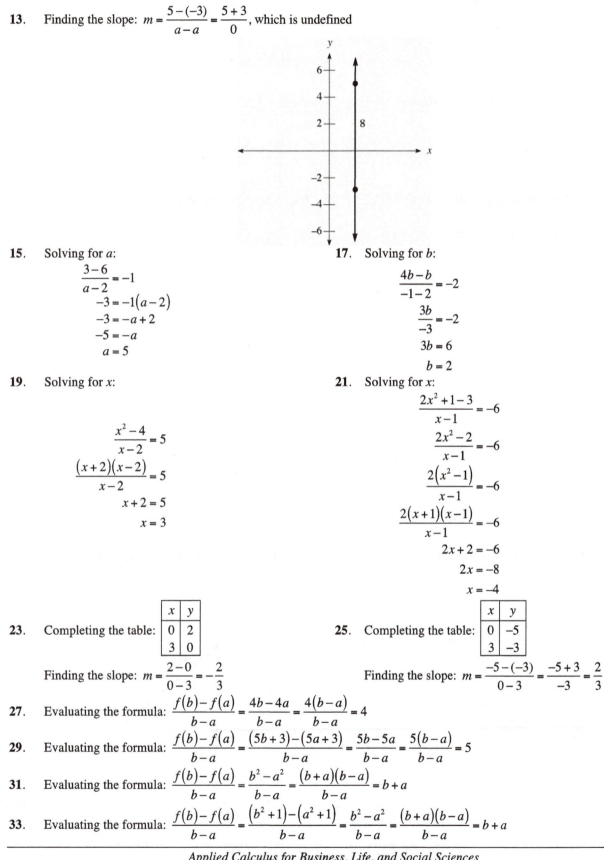

15. Solving for a:

$$\frac{3-6}{a-2} = -1$$
$$-3 = -1(a-2)$$
$$-3 = -a+2$$
$$-5 = -a$$
$$a = 5$$

17. Solving for b:

$$\frac{4b-b}{-1-2} = -2$$
$$\frac{3b}{-3} = -2$$
$$3b = 6$$
$$b = 2$$

19. Solving for x:

$$\frac{x^2-4}{x-2} = 5$$
$$\frac{(x+2)(x-2)}{x-2} = 5$$
$$x+2 = 5$$
$$x = 3$$

21. Solving for x:

$$\frac{2x^2+1-3}{x-1} = -6$$
$$\frac{2x^2-2}{x-1} = -6$$
$$\frac{2(x^2-1)}{x-1} = -6$$
$$\frac{2(x+1)(x-1)}{x-1} = -6$$
$$2x+2 = -6$$
$$2x = -8$$
$$x = -4$$

23. Completing the table:

x	y
0	2
3	0

Finding the slope: $m = \dfrac{2-0}{0-3} = -\dfrac{2}{3}$

25. Completing the table:

x	y
0	-5
3	-3

Finding the slope: $m = \dfrac{-5-(-3)}{0-3} = \dfrac{-5+3}{-3} = \dfrac{2}{3}$

27. Evaluating the formula: $\dfrac{f(b)-f(a)}{b-a} = \dfrac{4b-4a}{b-a} = \dfrac{4(b-a)}{b-a} = 4$

29. Evaluating the formula: $\dfrac{f(b)-f(a)}{b-a} = \dfrac{(5b+3)-(5a+3)}{b-a} = \dfrac{5b-5a}{b-a} = \dfrac{5(b-a)}{b-a} = 5$

31. Evaluating the formula: $\dfrac{f(b)-f(a)}{b-a} = \dfrac{b^2-a^2}{b-a} = \dfrac{(b+a)(b-a)}{b-a} = b+a$

33. Evaluating the formula: $\dfrac{f(b)-f(a)}{b-a} = \dfrac{(b^2+1)-(a^2+1)}{b-a} = \dfrac{b^2-a^2}{b-a} = \dfrac{(b+a)(b-a)}{b-a} = b+a$

35. Evaluating the formula:

$$\frac{f(b)-f(a)}{b-a}=\frac{\left(b^2-3b+4\right)-\left(a^2-3a+4\right)}{b-a}$$

$$=\frac{b^2-a^2-3b+3a}{b-a}$$

$$=\frac{(b+a)(b-a)-3(b-a)}{b-a}$$

$$=\frac{(b-a)(b+a-3)}{b-a}$$

$$=b+a-3$$

37. Using the slope-intercept formula: $y=-4x-3$

39. Using the slope-intercept formula: $y=-\frac{2}{3}x$

41. Using the slope-intercept formula: $y=-\frac{2}{3}x+\frac{1}{4}$

43. The slope is 3 and the y-intercept is –2:

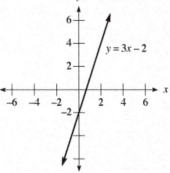

45. Solving for y:

$$2x-3y=12$$

$$-3y=-2x+12$$

$$y=\frac{2}{3}x-4$$

The slope is $\frac{2}{3}$ and the y-intercept is –4:

47. Solving for y:

$$4x+5y=20$$

$$5y=-4x+20$$

$$y=-\frac{4}{5}x+4$$

The slope is $-\frac{4}{5}$ and the y-intercept is 4:

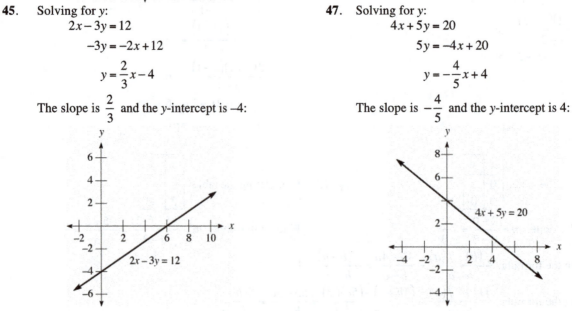

49. Using the point-slope formula:

$$y-(-5)=2(x-(-2))$$
$$y+5=2(x+2)$$
$$y+5=2x+4$$
$$y=2x-1$$

51. Using the point-slope formula:

$$y-1=-\frac{1}{2}(x-(-4))$$
$$y-1=-\frac{1}{2}(x+4)$$
$$y-1=-\frac{1}{2}x-2$$
$$y=-\frac{1}{2}x-1$$

53. Using the point-slope formula:

$$y-2=-3\left(x-\left(-\frac{1}{3}\right)\right)$$
$$y-2=-3\left(x+\frac{1}{3}\right)$$
$$y-2=-3x-1$$
$$y=-3x+1$$

55. Using the point-slope formula:

$$y-2=\frac{2}{3}(x-(-4))$$
$$y-2=\frac{2}{3}(x+4)$$
$$y-2=\frac{2}{3}x+\frac{8}{3}$$
$$y=\frac{2}{3}x+\frac{14}{3}$$

57. Using the point-slope formula:

$$y-(-2)=-\frac{1}{4}(x-(-5))$$
$$y+2=-\frac{1}{4}(x+5)$$
$$y+2=-\frac{1}{4}x-\frac{5}{4}$$
$$y=-\frac{1}{4}x-\frac{13}{4}$$

59. First find the slope: $m=\dfrac{1-(-2)}{-2-3}=\dfrac{1+2}{-5}=-\dfrac{3}{5}$. Using the point-slope formula:

$$y-(-2)=-\frac{3}{5}(x-3)$$
$$5(y+2)=-3(x-3)$$
$$5y+10=-3x+9$$
$$5y=-3x-1$$
$$y=-\frac{3}{5}x-\frac{1}{5}$$

61. First find the slope: $m=\dfrac{\frac{1}{3}-\frac{1}{2}}{-4-(-2)}=\dfrac{-\frac{1}{6}}{-4+2}=\dfrac{-\frac{1}{6}}{-2}=\dfrac{1}{12}$. Using the point-slope formula:

$$y-\frac{1}{2}=\frac{1}{12}(x-(-2))$$
$$12\left(y-\frac{1}{2}\right)=1(x+2)$$
$$12y-6=x+2$$
$$12y=x+8$$
$$y=\frac{1}{12}x+\frac{2}{3}$$

63. First find the slope: $m = \dfrac{-1-\left(-\dfrac{1}{5}\right)}{-\dfrac{1}{3}-\dfrac{1}{3}} = \dfrac{-1+\dfrac{1}{5}}{-\dfrac{2}{3}} = \dfrac{-\dfrac{4}{5}}{-\dfrac{2}{3}} = \dfrac{4}{5} \cdot \dfrac{3}{2} = \dfrac{6}{5}$. Using the point-slope formula:

$$y-(-1) = \frac{6}{5}\left(x-\left(-\frac{1}{3}\right)\right)$$

$$y+1 = \frac{6}{5}\left(x+\frac{1}{3}\right)$$

$$5(y+1) = 6\left(x+\frac{1}{3}\right)$$

$$5y+5 = 6x+2$$

$$5y = 6x-3$$

$$y = \frac{6}{5}x - \frac{3}{5}$$

65. **a.** For the x-intercept, substitute $y = 0$:

$$3x-2(0) = 10$$
$$3x = 10$$
$$x = \frac{10}{3}$$

The x-intercept is $\left(\dfrac{10}{3}, 0\right)$.

For the y-intercept, substitute $x = 0$:

$$3(0)-2y = 10$$
$$-2y = 10$$
$$y = -5$$

The y-intercept is $(0,-5)$.

b. Substituting $y = 1$:

$$3x-2(1) = 10$$
$$3x-2 = 10$$
$$3x = 12$$
$$x = 4$$

Another solution is (4,1). Other answers are possible.

c. Solving for y:

$$3x-2y = 10$$
$$-2y = -3x+10$$
$$y = \frac{3}{2}x-5$$

d. Substituting $x = 2$: $y = \dfrac{3}{2}(2)-5 = 3-5 = -2$.

No, the point (2,2) is not a solution to the equation.

67. **a.** Solving for x:

$$-2x+1 = -3$$
$$-2x = -4$$
$$x = 2$$

b. Simplifying: $-2x+1-3 = -2x-2$

c. Substituting $y = 0$:

$$-2x+0 = -3$$
$$-2x = -3$$
$$x = \frac{3}{2}$$

d. Substituting $x = 0$: $y = 2(0)-3 = -3$

e.   Sketching the graph:

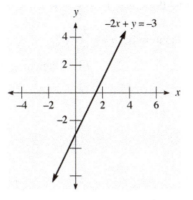

f.   Solving for y:
$$-2x + y = -3$$
$$y = 2x - 3$$

69.   **a.**   The slope is $\dfrac{1}{2}$, the x-intercept is $(0,0)$, and the y-intercept is $(0,0)$:

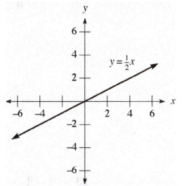

b.   There is no slope, the x-intercept is $(3,0)$, and there is no y-intercept:

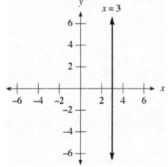

c.   The slope is 0, there is no x-intercept, and the y-intercept is $(0,-2)$:

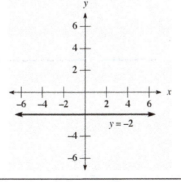

71. Using the points (3,0) and (0,2), first find the slope: $m = \dfrac{2-0}{0-3} = -\dfrac{2}{3}$

Using the slope-intercept formula, the equation is: $y = -\dfrac{2}{3}x + 2$

73. Sketching the graph:

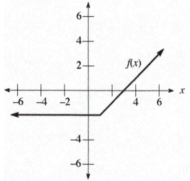
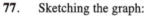

75. Sketching the graph:

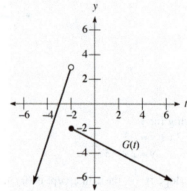

77. Sketching the graph:

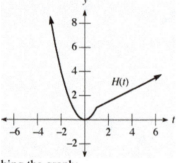

79. Sketching the graph:

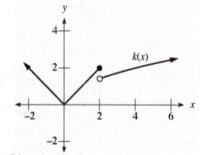
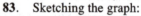

81. Sketching the graph:

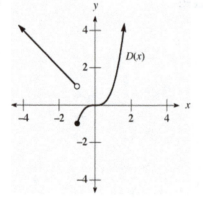

83. Sketching the graph:

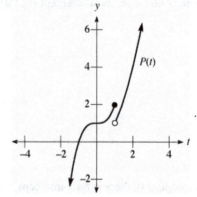

85. Writing in piecewise form:

$$f(x) = \begin{cases} -2x + 8 & \text{if } x < 4 \\ 2x - 8 & \text{if } x \geq 4 \end{cases}$$

Sketching the graph:

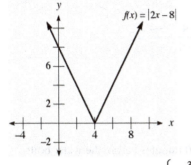

87. Writing in piecewise form:

$$g(t) = \begin{cases} -\dfrac{t}{3} - 1 & \text{if } t < -3 \\ \dfrac{t}{3} + 1 & \text{if } t \geq -3 \end{cases}$$

Sketching the graph:

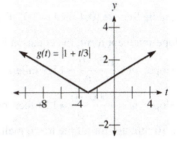

89. Writing in piecewise form: $F(x) = \begin{cases} -x^3 & \text{if } x < 0 \\ x^3 & \text{if } x \geq 0 \end{cases}$

Sketching the graph:

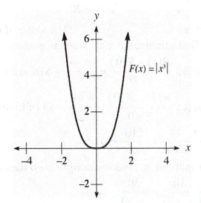

91. Graphing the line:

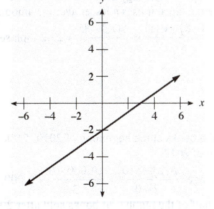

The slope is $m = \dfrac{2}{3}$.

93. The slope is $\dfrac{3}{2}$.

95. There is no slope (undefined).

97. The slope is $\dfrac{2}{3}$.

99. The slope is $\frac{1}{2}$ and the y-intercept is –4. Using the slope-intercept form, the equation is $y = \frac{1}{2}x - 4$.

101. The slope is $-\frac{2}{3}$ and the y-intercept is 3. Using the slope-intercept form, the equation is $y = -\frac{2}{3}x + 3$.

103. Two points on the line are (0,–4) and (2,0). Finding the slope: $m = \frac{0-(-4)}{2-0} = \frac{4}{2} = 2$

    Using the slope-intercept form, the equation is $y = 2x - 4$.

105. Two points on the line are (0,4) and (–2,0). Finding the slope: $m = \frac{0-4}{-2-0} = \frac{-4}{-2} = 2$

    Using the slope-intercept form, the equation is $y = 2x + 4$.

107. Finding the slope: $m = \frac{105-0}{6-0} = 17.5$ miles/hour

109. Finding the slope: $m = \frac{3600-0}{30-0} = 120$ feet/second

111. **a.** It takes 10 minutes for all the ice to melt. **b.** It takes 20 minutes before the water boils.

    c. A piecewise function is: $T(x) = \begin{cases} 20x - 20 & \text{if } 0 \le x < 1 \\ 0 & \text{if } 1 \le x < 10 \\ 10x - 100 & \text{if } 10 \le x < 20 \\ 0 & \text{if } x \ge 20 \end{cases}$

    d. The slope of A is 20°C per minute. **e.** The slope of C is 10°C per minute.
    f. It is changing faster during the first minute, since its slope is greater.

113. **a.** Finding the average rate of change: $\frac{R(1)-R(0)}{1-0} = \frac{16-0}{1} = \$16$ million/year

    b. Finding the average rate of change: $\frac{R(2)-R(0)}{2-0} = \frac{6-0}{2} = \$3$ million/year

115. **a.** Computing the slope: $m \approx \frac{150-40}{2600-450} = \frac{110}{2150} \approx 0.051$

    For each additional lumen of output, the incandescent light bulb uses an average of 0.051 watts of energy.

    b. Computing the slope: $m \approx \frac{40-10}{2600-450} = \frac{30}{2150} \approx 0.014$

    For each additional lumen of output, the energy efficient bulb uses an average of 0.014 watts of energy.

    c. The energy efficient bulb is better, since it uses a lesser average amount of energy per lumen of output.

117. Finding the average rate of change: $\frac{V(8)-V(1)}{8-1} \approx \frac{247.3-85}{7} \approx 23.2$ mph/second

119. **a.** Finding the values:

    $S(0) = 0$
    $S(1) = 5,000 \cdot 1 = 5,000$
    $S(2) = 5,000 \cdot 2 = 10,000$

    XYZ Textbooks/MathTV sold 0 books at the beginning of 2010, 5,000 books by the beginning of 2011, and 10,000 books by the beginning of 2012.

    b. Finding the average rate of change: $\frac{S(2)-S(0)}{2-0} = \frac{10,000-0}{2} = 5,000$ books/year

    The average rate of change is half of the number of books sold after 2 years.

    c. Finding the average rate of change: $\frac{S(3)-S(1)}{3-1} = \frac{20,000-5,000}{2} = 7,500$ books/year

121. a. Finding the equation:
$$E = 0.6(220 - A))$$
$$E = 132 - 0.6A$$

b. Substituting $A = 22$: $E = 132 - 0.6(22) = 132 - 13.2 \approx 119$ beats/minute

c. Sketching the graph:

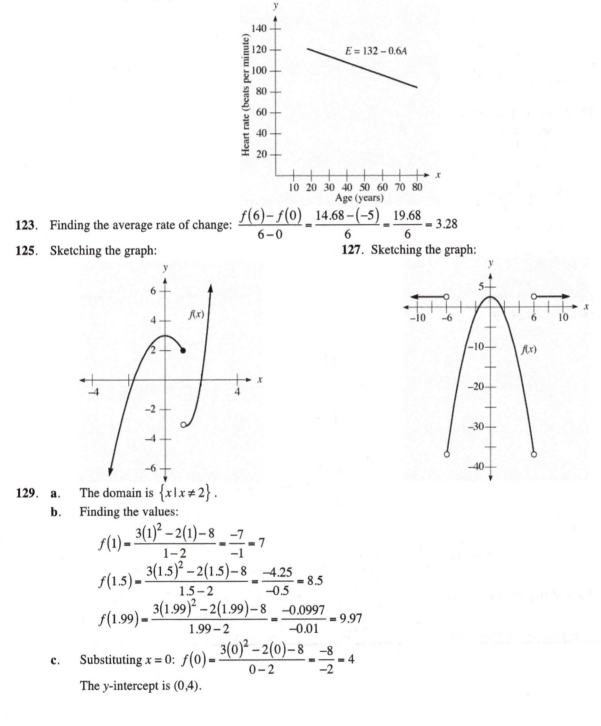

123. Finding the average rate of change: $\dfrac{f(6) - f(0)}{6 - 0} = \dfrac{14.68 - (-5)}{6} = \dfrac{19.68}{6} = 3.28$

125. Sketching the graph:

127. Sketching the graph:

129. a. The domain is $\{x \mid x \neq 2\}$.

b. Finding the values:
$$f(1) = \frac{3(1)^2 - 2(1) - 8}{1 - 2} = \frac{-7}{-1} = 7$$
$$f(1.5) = \frac{3(1.5)^2 - 2(1.5) - 8}{1.5 - 2} = \frac{-4.25}{-0.5} = 8.5$$
$$f(1.99) = \frac{3(1.99)^2 - 2(1.99) - 8}{1.99 - 2} = \frac{-0.0997}{-0.01} = 9.97$$

c. Substituting $x = 0$: $f(0) = \dfrac{3(0)^2 - 2(0) - 8}{0 - 2} = \dfrac{-8}{-2} = 4$

The y-intercept is $(0,4)$.

d. Substituting $y = 0$:

$$\frac{3x^2 - 2x - 8}{x - 2} = 0$$

$$\frac{(3x + 4)(x - 2)}{x - 2} = 0$$

$$3x + 4 = 0$$

$$3x = -4$$

$$x = -\frac{4}{3}$$

The x-intercept is $\left(-\frac{4}{3}, 0\right)$.

131. a. The domain is $\{x \mid x \neq 3\}$.

b. Finding the values:

$$f(2) = \frac{1}{(2 - 3)^2} = \frac{1}{1} = 1$$

$$f(5) = \frac{1}{(5 - 3)^2} = \frac{1}{4}$$

c. Substituting $y = 0$:

$$\frac{1}{(x - 3)^2} = 0$$

$$1 \neq 0$$

There is no x-intercept.

d. Sketching the graph:

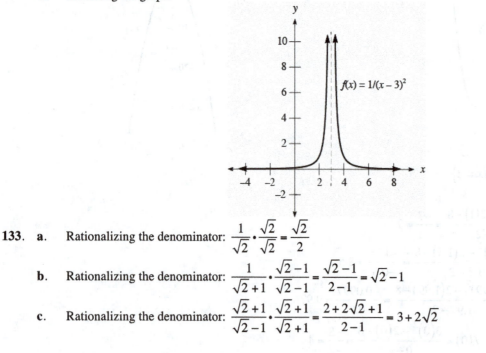

133. a. Rationalizing the denominator: $\dfrac{1}{\sqrt{2}} \cdot \dfrac{\sqrt{2}}{\sqrt{2}} = \dfrac{\sqrt{2}}{2}$

b. Rationalizing the denominator: $\dfrac{1}{\sqrt{2} + 1} \cdot \dfrac{\sqrt{2} - 1}{\sqrt{2} - 1} = \dfrac{\sqrt{2} - 1}{2 - 1} = \sqrt{2} - 1$

c. Rationalizing the denominator: $\dfrac{\sqrt{2} + 1}{\sqrt{2} - 1} \cdot \dfrac{\sqrt{2} + 1}{\sqrt{2} + 1} = \dfrac{2 + 2\sqrt{2} + 1}{2 - 1} = 3 + 2\sqrt{2}$

1.4 Introduction to Limits

1. Finding the limit: $\lim\limits_{x\to a} f(x) = P$

3. **a.** Finding the limit: $\lim\limits_{x\to 10} f(x) = 6$ **b.** Finding the limit: $\lim\limits_{x\to 3} f(x) = 4$

5. Finding the limit: $\lim\limits_{x\to 8} f(x) = 13$ since $\lim\limits_{x\to 8^-} f(x) = 13$ and $\lim\limits_{x\to 8^+} f(x) = 13$, which are equal.

7. Finding the limit: $\lim\limits_{x\to 6} f(x)$ does not exist since $\lim\limits_{x\to 6^-} f(x) = 2$ and $\lim\limits_{x\to 6^+} f(x) = 4$, which are not equal.

9. Completing the tables:

x (from the left)	$f(x)$
0.5	−5.75
0.7	−4.71
0.9	−3.59
0.99	−3.0599
0.999	−3.005999

x (from the right)	$f(x)$
1.5	0.25
1.3	−1.11
1.1	−2.39
1.01	−2.9399
1.001	−2.993999

Thus $\lim\limits_{x\to 1}\left(x^2 + 4x - 8\right) = -3$.

11. Finding the limit: $\lim\limits_{x\to 3} 5 = 5$

13. Finding the limit: $\lim\limits_{x\to 2} x = 2$

15. Finding the limit: $\lim\limits_{x\to -1} 3x = 3(-1) = -3$

17. Finding the limit: $\lim\limits_{x\to -2}(5x - 1) = -10 - 1 = -11$

19. Finding the limit: $\lim\limits_{x\to 0}\left(x^2 - 4\right) = 0 - 4 = -4$

21. Finding the limit: $\lim\limits_{x\to 2}\dfrac{x-3}{x+1} = -\dfrac{1}{3}$

23. Finding the limit: $\lim\limits_{x\to -4}\dfrac{x-5}{x+4}$ does not exist

25. Finding the limit: $\lim\limits_{x\to\infty}\dfrac{2x-5}{3x+1} = \lim\limits_{x\to\infty}\dfrac{2x}{3x} = \lim\limits_{x\to\infty}\dfrac{2}{3} = \dfrac{2}{3}$

27. Finding the limit: $\lim\limits_{x\to\infty}\dfrac{7x-3}{x^3+2} = \lim\limits_{x\to\infty}\dfrac{7x}{x^3} = \lim\limits_{x\to\infty}\dfrac{7}{x^2} = 0$

29. Finding the limit: $\lim\limits_{x\to\infty}\dfrac{16x^3-3x+7}{9x^2-5x} = \lim\limits_{x\to\infty}\dfrac{16x^3}{9x^2} = \lim\limits_{x\to\infty}\dfrac{16x}{9} = \infty$, which does not exist

31. Finding the limit: $\lim\limits_{x\to\infty}\dfrac{(5x-3)(x+2)}{x^2-3} = \lim\limits_{x\to\infty}\dfrac{5x^2+7x-6}{x^2-3} = \lim\limits_{x\to\infty}\dfrac{5x^2}{x^2} = \lim\limits_{x\to\infty} 5 = 5$

33. Finding the limit: $\lim\limits_{t\to\infty}\dfrac{850t^2}{t^2+20} = \lim\limits_{t\to\infty}\dfrac{850t^2}{t^2} = \lim\limits_{t\to\infty} 850 = 850$

35. Finding the limit: $\lim\limits_{x\to\infty} f(x) = 1$

37. **a.** Finding the limit: $\lim\limits_{x\to\infty} f(x) = -1$

     b. Finding the limit: $\lim\limits_{x\to -\infty} f(x) = -1$

     c. Finding the limit: $\lim\limits_{x\to 2} f(x) = \pm\infty$, which does not exist

39. **a.** Finding the limit: $\lim\limits_{x\to\infty} f(x) = \infty$, which does not exist

     b. Finding the limit: $\lim\limits_{x\to -\infty} f(x) = \infty$, which does not exist

     c. Finding the limit: $\lim\limits_{x\to 2} f(x) = 4$

41. **a.** As x approaches -2, y approaches 2.

    b. Finding the limit: $\lim\limits_{x \to 0} |x| = 0$

    c. This statement is false, since $\lim\limits_{x \to 2} |x| = 2$.

43. **a.** As x approaches $\dfrac{1}{2}$, $f(x)$ approaches 2.

    b. The value is $a = -1$, since $\lim\limits_{x \to -1} \dfrac{1}{x} = -1$

    c. This statement is true.

45. **a.** Finding the limit: $\lim\limits_{x \to 210} C(x)$ does not exist, since $\lim\limits_{x \to 210^-} 0.05x = 10.5$ and $\lim\limits_{x \to 210^+} 0.08x = 16.8$. The monthly charge for electricity approaches \$10.50 as usage increases to 210 kilowatt hours, while the monthly charge for electricity approaches \$16.80 as usage decreases to 210 kilowatt hours.

    b. Finding the limit: $\lim\limits_{x \to 340} C(x) = 27.20$, since $\lim\limits_{x \to 340^-} 0.08x = 27.20$ and $\lim\limits_{x \to 340^+} (0.1x - 6.8) = 27.20$. The monthly charge for electricity approaches \$27.20 as usage approaches 340 kilowatt hours (from either direction).

    c. Finding the limit: $\lim\limits_{x \to 450} C(x) = 38.20$, since $\lim\limits_{x \to 450^-} (0.1x - 6.8) = 38.20$ and $\lim\limits_{x \to 450^+} (0.12x - 15.8) = 38.20$.

    The monthly charge for electricity approaches \$38.20 as usage approaches 450 kilowatt hours (from either direction).

47. Finding the limit: $\lim\limits_{x \to \infty} \overline{C}(x) = \lim\limits_{x \to \infty} \left(0.89 + \dfrac{1,500}{x} \right) = 0.89$

As the production level (x) increases, the average cost approaches \$0.89 per tube.

49. Finding the limit: $\lim\limits_{p \to \infty} N(p) = \lim\limits_{p \to \infty} \dfrac{500}{1 + p} = \lim\limits_{p \to \infty} \dfrac{500}{p} = 0$

If the level of PCB increases without control, the number of fish will decrease to 0.

51. **a.** Evaluating the function: $\overline{C}(100) = \dfrac{600 + 8,000}{200 + 300} = \dfrac{8,600}{500} = \$17.20$ per box

    b. Finding the limit: $\lim\limits_{x \to \infty} \overline{C}(x) = \lim\limits_{x \to \infty} \dfrac{6x + 8,000}{2x + 300} = \lim\limits_{x \to \infty} \dfrac{6x}{2x} = \lim\limits_{x \to \infty} \dfrac{6}{2} = \$3.00$ per box

53. Finding the limit: $\lim\limits_{t \to \infty} C(t) = \lim\limits_{t \to \infty} \dfrac{36t}{t^2 + 12} = \lim\limits_{t \to \infty} \dfrac{36t}{t^2} = \lim\limits_{t \to \infty} \dfrac{36}{t} = 0$

In the long run, the concentration will decrease to 0 milligrams per liter.

55. The vertical asymptote is $t = -3$.

Finding the limit: $\lim\limits_{t \to \infty} V(t) = \lim\limits_{t \to \infty} \dfrac{340t}{t + 3} = \lim\limits_{t \to \infty} \dfrac{340t}{t} = \lim\limits_{t \to \infty} 340 = 340$

The horizontal asymptote is $y = 340$.

57. **a.** Evaluating: $g(2) = 3$

    b. Finding the limit: $\lim\limits_{x \to 2^-} g(x) = \lim\limits_{x \to 2^-} \dfrac{x^2 - x - 2}{x - 2} = \lim\limits_{x \to 2^-} \dfrac{(x-2)(x+1)}{x - 2} = \lim\limits_{x \to 2^-} (x + 1) = 3$

    c. Finding the limit: $\lim\limits_{x \to 2^+} g(x) = \lim\limits_{x \to 2^+} \dfrac{x^2 - x - 2}{x - 2} = \lim\limits_{x \to 2^+} \dfrac{(x-2)(x+1)}{x - 2} = \lim\limits_{x \to 2^+} (x + 1) = 3$

    d. Since $\lim\limits_{x \to 2^-} g(x) = 3$ and $\lim\limits_{x \to 2^+} g(x) = 3$, $\lim\limits_{x \to 2} g(x) = 3$.

1.5 Functions and Continuity

1. Since $\lim\limits_{x \to -1} f(x) = \lim\limits_{x \to -1}\left(3x^2 - 7x + 4\right) = 3(-1)^2 - 7(-1) + 4 = 14 = f(-1)$, the function is continuous at $x = -1$.

3. Since $\lim\limits_{x \to 3} f(x) = \lim\limits_{x \to 3} \dfrac{x^2 - 5x + 6}{x - 2} = \lim\limits_{x \to 3} \dfrac{(x-2)(x-3)}{x-2} = \lim\limits_{x \to 3}(x-3) = 0 = f(3)$, the function is continuous at $x = 3$.

5. Since $\lim\limits_{x \to 2} f(x) = \lim\limits_{x \to 2} \dfrac{x^2 - 5x + 6}{x - 2} = \lim\limits_{x \to 2} \dfrac{(x-2)(x-3)}{x-2} = \lim\limits_{x \to 2}(x-3) = -1$ but $f(2)$ is undefined, the function is discontinuous at $x = 2$.

7. Since $f(1)$ is undefined, the function is discontinuous at $x = 1$.

9. Since $\lim\limits_{x \to 0} f(x) = \lim\limits_{x \to 0}|x - 3| = |0 - 3| = 3 = f(0)$, the function is continuous at $x = 0$.

11. Since $\lim\limits_{x \to -4^-} f(x) = \lim\limits_{x \to -4^-}\left(x^2 + 8x + 13\right) = (-4)^2 + 8(-4) + 13 = -3$ and $\lim\limits_{x \to -4^+} f(x) = \lim\limits_{x \to -4^+}(-3) = -3$,
$\lim\limits_{x \to -4} f(x) = -3$. Since $f(-4) = 0$, the function is continuous at $x = -4$.

13. Since $\lim\limits_{x \to 3} f(x) = \lim\limits_{x \to 3} \dfrac{2x^2 - 5x - 3}{x - 3} = \lim\limits_{x \to 3} \dfrac{(x-3)(2x+1)}{x-3} = \lim\limits_{x \to 3}(2x+1) = 7 = f(3)$, the function is continuous at $x = 3$.

15. The function is discontinuous at $x = a$, since $\lim\limits_{x \to a^-} f(x) \neq \lim\limits_{x \to a^+} f(x)$ and thus $\lim\limits_{x \to a} f(x)$ does not exist.

17. The function is discontinuous at $x = a$, since $f(a)$ does not exist.

19. The function is continuous at $x = a$ since all three continuity conditions are met.

21. The function is continuous at $x = a$ since all three continuity conditions are met.

23. The function is discontinuous at $x = a$, since $\lim\limits_{x \to a} f(x) \neq f(a)$.

25. The function is discontinuous at $x = a$, since $\lim\limits_{x \to a^-} f(x) \neq \lim\limits_{x \to a^+} f(x)$ and thus $\lim\limits_{x \to a} f(x)$ does not exist.

27. The function is discontinuous at $t = a$, since $\lim\limits_{t \to a^-} A(t) \neq \lim\limits_{t \to a^+} A(t)$ and thus $\lim\limits_{t \to a} A(t)$ does not exist. A possible interpretation is that, at $t = a$, a conceptual leap in learning ("aha" moment) exists.

29. The function is discontinuous at $t = a$, since $\lim\limits_{t \to a^-} P(t) \neq \lim\limits_{t \to a^+} P(t)$ and thus $\lim\limits_{t \to a} P(t)$ does not exist. A possible interpretation is that, at $t = a$, a drastic reduction in profits occurred (such as an economic recession, or a change in leadership). At $t = b$ the profits have been reduced to 0.

31. The function is discontinuous at $t = a$, since $\lim\limits_{t \to a^-} P(t) \neq \lim\limits_{t \to a^+} P(t)$ and thus $\lim\limits_{t \to a} P(t)$ does not exist. A possible interpretation is that, at $t = a$, a drastic increase in pollution occurred (such as a ruptured pipeline).

33. Answers will vary.

35. a. The domain of this function is $\{x \mid x \geq 0\}$.

    b. The function is discontinuous at $x = -150$. Since this value is not in the domain, it does not appear on the graph.

37. The function is discontinuous at $x = -4$:

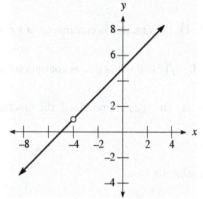

39. The function is discontinuous at $x = -4, x = -1$:

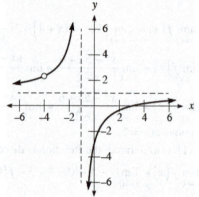

41. The function is continuous everywhere:

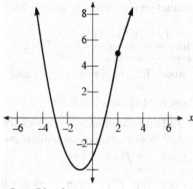

43. Simplifying: $f(x+h) = 5(x+h) + 8 = 5x + 5h + 8$

45. Simplifying:

$$\frac{f(x+h)-f(x)}{h} = \frac{\left(-(x+h)^2 + 13(x+h) - 22\right) - \left(-x^2 + 13x - 22\right)}{h}$$

$$= \frac{-x^2 - 2xh - h^2 + 13x + 13h - 22 + x^2 - 13x + 22}{h}$$

$$= \frac{-2xh - h^2 + 13h}{h}$$

$$= -2x - h + 13$$

47. Finding the limit:

$$\lim_{h \to 0} \frac{f(x+h)-f(x)}{h} = \lim_{h \to 0} \frac{\left(-5(x+h)^2 + 200(x+h)\right) - \left(-5x^2 + 200x\right)}{h}$$

$$= \lim_{h \to 0} \frac{-5x^2 - 10xh - 5h^2 + 200x + 200h + 5x^2 - 200x}{h}$$

$$= \lim_{h \to 0} \frac{-10xh - 5h^2 + 200h}{h}$$

$$= \lim_{h \to 0} (-10x - 5h + 200)$$

$$= -10x + 200$$

1.6 Average and Instantaneous Rates of Change

1. Finding the average rate of change: $\dfrac{f(4)-f(1)}{4-1} = \dfrac{\left(4^2-2\right)-\left(1^2-2\right)}{3} = \dfrac{14+1}{3} = \dfrac{15}{3} = 5$

3. Finding the average rate of change: $\dfrac{f(5)-f(-1)}{5-(-1)} = \dfrac{(4\cdot 5-5)-(4\cdot(-1)-5)}{6} = \dfrac{15+9}{6} = \dfrac{24}{6} = 4$

5. Finding the average rate of change: $\dfrac{f(15)-f(6)}{15-6} = \dfrac{\dfrac{15^2-25}{15-5}-\dfrac{6^2-25}{6-5}}{9} = \dfrac{20-11}{9} = \dfrac{9}{9} = 1$

7. Finding the difference quotient: $\dfrac{f(x+h)-f(x)}{h} = \dfrac{(3(x+h)-8)-(3x-8)}{h} = \dfrac{3x+3h-8-3x+8}{h} = \dfrac{3h}{h} = 3$

9. Finding the difference quotient:

$$\dfrac{f(x+h)-f(x)}{h} = \dfrac{\left(-\sqrt{x+h}+2\right)-\left(-\sqrt{x}+2\right)}{h}$$

$$= \dfrac{-\sqrt{x+h}+\sqrt{x}}{h}$$

$$= \dfrac{-\sqrt{x+h}+\sqrt{x}}{h} \cdot \dfrac{-\sqrt{x+h}-\sqrt{x}}{-\sqrt{x+h}-\sqrt{x}}$$

$$= \dfrac{x+h-x}{-h\left(\sqrt{x+h}+\sqrt{x}\right)}$$

$$= \dfrac{h}{-h\left(\sqrt{x+h}+\sqrt{x}\right)}$$

$$= \dfrac{-1}{\sqrt{x+h}+\sqrt{x}}$$

11. Finding the difference quotient:

$$\dfrac{f(x+h)-f(x)}{h} = \dfrac{-\dfrac{1}{x+h}+\dfrac{1}{x}}{h} = \dfrac{-\dfrac{1}{x+h}+\dfrac{1}{x}}{h} \cdot \dfrac{x(x+h)}{x(x+h)} = \dfrac{-x+x+h}{hx(x+h)} = \dfrac{h}{hx(x+h)} = \dfrac{1}{x(x+h)}$$

13. Finding the instantaneous rate of change:

$$\lim_{h\to 0} \dfrac{f(x+h)-f(x)}{h} = \lim_{h\to 0} \dfrac{\left((x+h)^2+4\right)-\left(x^2+4\right)}{h}$$

$$= \lim_{h\to 0} \dfrac{x^2+2xh+h^2+4-x^2-4}{h}$$

$$= \lim_{h\to 0} \dfrac{2xh+h^2}{h}$$

$$= \lim_{h\to 0} (2x+h)$$

$$= 2x$$

Evaluating when $x = 3$: $2(3) = 6$

---

15. Finding the instantaneous rate of change:

$$\lim_{h \to 0} \frac{f(x+h)-f(x)}{h} = \lim_{h \to 0} \frac{\left(5(x+h)^2 + 4(x+h)+2\right)-\left(5x^2+4x+2\right)}{h}$$

$$= \lim_{h \to 0} \frac{5x^2+10xh+5h^2+4x+4h+2-5x^2-4x-2}{h}$$

$$= \lim_{h \to 0} \frac{10xh+5h^2+4h}{h}$$

$$= \lim_{h \to 0} (10x+5h+4)$$

$$= 10x+4$$

Evaluating when $x = 0$: $10(0)+4 = 0+4 = 4$

17. Finding the instantaneous rate of change:

$$\lim_{h \to 0} \frac{f(x+h)-f(x)}{h} = \lim_{h \to 0} \frac{\left(4(x+h)+4\right)-\left(4x+4\right)}{h} = \lim_{h \to 0} \frac{4x+4h+4-4x-4}{h} = \lim_{h \to 0} \frac{4h}{h} = \lim_{h \to 0} 4 = 4$$

Evaluating when $x = -2$: 4

19. Finding the instantaneous rate of change:

$$\lim_{h \to 0} \frac{f(x+h)-f(x)}{h} = \lim_{h \to 0} \frac{\dfrac{x+h+2}{x+h-3} - \dfrac{x+2}{x-3}}{h}$$

$$= \lim_{h \to 0} \frac{\dfrac{x+h+2}{x+h-3} - \dfrac{x+2}{x-3}}{h} \cdot \frac{(x+h-3)(x-3)}{(x+h-3)(x-3)}$$

$$= \lim_{h \to 0} \frac{(x+h+2)(x-3)-(x+2)(x+h-3)}{h(x+h-3)(x-3)}$$

$$= \lim_{h \to 0} \frac{\left(x^2+xh-x-3h-6\right)-\left(x^2+xh-x+2h-6\right)}{h(x+h-3)(x-3)}$$

$$= \lim_{h \to 0} \frac{-5h}{h(x+h-3)(x-3)}$$

$$= \lim_{h \to 0} \frac{-5}{(x+h-3)(x-3)}$$

$$= \frac{-5}{(x-3)^2}$$

Evaluating when $x = 2$: $\dfrac{-5}{(2-3)^2} = \dfrac{-5}{1} = -5$

21. Evaluating: $f(20)-f(0) = \left(1{,}200 \cdot 20 - 20^3\right)-\left(1{,}200 \cdot 0 - 0^3\right) = 16{,}000-0 = 16{,}000$

23. Evaluating: $f(5)-f(2) = \left(\dfrac{5^3}{3} - \dfrac{5 \cdot 5^2}{2} + 4 \cdot 5\right) - \left(\dfrac{2^3}{3} - \dfrac{5 \cdot 2^2}{2} + 4 \cdot 2\right) = -\dfrac{5}{6} - \dfrac{2}{3} = -\dfrac{3}{2}$

25. Finding the average rate of change: $\dfrac{C(8)-C(0)}{8-0} = \dfrac{0.1-0.8}{8-0} = \dfrac{-0.7}{8} = -0.0875$

The drug concentration decreases by an average of 0.0875 mg/cc per hour during the first 8 hours after injection.

27. The profit increases by an average of 60 cents per cent increase in the selling price from 75 to 80 cents.

29. Evaluating when $x = 4$: $8(4)-1 = 32-1 = 31$

The function (y) increases by 31 at the instant when $x = 4$.

31. Finding the average rate of change: $\dfrac{62{,}000-45{,}200}{2020-2010}=\dfrac{16{,}800}{10}=1{,}680$

The physician assistant jobs will increase by an average of 1,680 jobs per year from 2010 to 2020.

33. **a.** Finding the average rate of change:

$$\frac{P(11)-P(10)}{11-10}=\frac{\left(-11^2+38\cdot 11-240\right)-\left(-10^2+38\cdot 10-240\right)}{11-10}=\frac{57-40}{1}=17$$

The average rate of change in weekly profit is $17 per fan.

b. Finding the average rate of change:

$$\frac{P(21)-P(20)}{21-20}=\frac{\left(-21^2+38\cdot 21-240\right)-\left(-20^2+38\cdot 20-240\right)}{21-20}=\frac{117-120}{1}=-3$$

The average rate of change in weekly profit is –$3 per fan.

35. **a.** Finding the change: $A(26)-A(25)=(0.08\cdot 26+19.7)-(0.08\cdot 25+19.7)=21.78-21.70=0.08$ years

b. Finding the change: $A(41)-A(40)=(0.08\cdot 41+19.7)-(0.08\cdot 40+19.7)=22.98-22.90=0.08$ years

c. Finding the change: $A(63)-A(62)=(0.08\cdot 63+19.7)-(0.08\cdot 62+19.7)=24.74-24.66=0.08$ years

37. Finding the instantaneous rate of change:

$$\lim_{h\to 0}\frac{V(t+h)-V(t)}{h}=\lim_{h\to 0}\frac{\dfrac{340(t+h)}{(t+h)+3}-\dfrac{340t}{t+3}}{h}$$

$$=\lim_{h\to 0}\frac{\dfrac{340t+340h}{t+h+3}-\dfrac{340t}{t+3}}{h}\cdot\frac{(t+h+3)(t+3)}{(t+h+3)(t+3)}$$

$$=\lim_{h\to 0}\frac{(340t+340h)(t+3)-(340t)(t+h+3)}{h(t+h+3)(t+3)}$$

$$=\lim_{h\to 0}\frac{\left(340t^2+340th+1020t+1020h\right)-\left(340t^2+340th+1020t\right)}{h(t+h+3)(t+3)}$$

$$=\lim_{h\to 0}\frac{1020h}{h(t+h+3)(t+3)}$$

$$=\lim_{h\to 0}\frac{1020}{(t+h+3)(t+3)}$$

$$=\frac{1020}{(t+3)^2}$$

Evaluating when $t=2$: $\dfrac{1020}{(2+3)^2}=\dfrac{1020}{25}=40.8$ mph per second

39. **a.** Finding the average rate of change: $\dfrac{f(15)-f(12)}{15-12}=\dfrac{(2\cdot 15+3)^{3/4}-(2\cdot 12+3)^{3/4}}{15-12}=\dfrac{33^{0.75}-27^{0.75}}{3}\approx 0.6413$

b. Using $h=0.001$: $\dfrac{f(12.001)-f(12)}{12.001-12}=\dfrac{(2\cdot 12.001+3)^{3/4}-(2\cdot 12+3)^{3/4}}{12.001-12}=\dfrac{27.002^{0.75}-27^{0.75}}{0.001}\approx 0.6580$

41. Evaluating: $f(5)=5^2+6\cdot 5-2=25+30-2=53$

43. Finding the limit:

$$\lim_{h \to 0} \frac{f(x+h)-f(x)}{h} = \lim_{h \to 0} \frac{\left((x+h)^2 + 6(x+h)-2\right)-\left(x^2+6x-2\right)}{h}$$

$$= \lim_{h \to 0} \frac{x^2+2xh+h^2+6x+6h-2-x^2-6x+2}{h}$$

$$= \lim_{h \to 0} \frac{2xh+h^2+6h}{h}$$

$$= \lim_{h \to 0} (2x+h+6)$$

$$= 2x+6$$

45. Using the point-slope formula:

$$y-1 = 4(x+1)$$
$$y-1 = 4x+4$$
$$y = 4x+5$$

47. Evaluating when $x = 8$: $f(8) = 6(8)^{-1/3} = 6(2)^{-1} = 6 \cdot \frac{1}{2} = 3$

Chapter 1 Test

1. Setting the denominator equal to 0:

$$x^2 - 9 = 0$$
$$(x+3)(x-3) = 0$$
$$x = -3, 3$$

The domain is $\{x \mid x \neq -3, 3\}$.

2.   **a.** Evaluating: $fg(3) = f(3) \cdot g(3) = (3-2)(3 \cdot 3+4) = (1)(13) = 13$

    b. Evaluating: $h(3) = 3(3)^2 - 2(3) - 8 = 27 - 6 - 8 = 13$

    c. Evaluating: $f(g(3)) = f(3 \cdot 3+4) = f(13) = 13 - 2 = 11$

3. The x-intercept is 3, the y-intercept is 6, and the slope is –2.

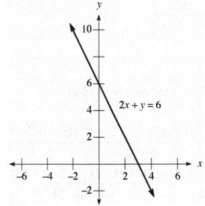

4. Using the point-slope formula:

$$y-3 = 2(x+1)$$
$$y-3 = 2x+2$$
$$y = 2x+5$$

5. First find the slope: $m = \dfrac{-1-2}{4-(-3)} = \dfrac{-3}{4+3} = -\dfrac{3}{7}$. Using the point-slope formula:

$$y - 2 = -\frac{3}{7}(x+3)$$
$$y - 2 = -\frac{3}{7}x - \frac{9}{7}$$
$$y = -\frac{3}{7}x + \frac{5}{7}$$

6. Graphing the function:

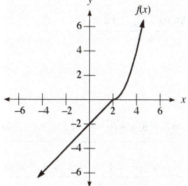

7. Finding the limit: $\lim\limits_{x \to 13} f(x) = 2$ since $\lim\limits_{x \to 13^-} f(x) = 2$ and $\lim\limits_{x \to 13^+} f(x) = 2$, which are equal.

8. **a.** Finding the limit: $\lim\limits_{x \to 1} f(x) = 4$

    b. Finding the limit: $\lim\limits_{x \to 3} f(x) = 6$

9. Finding the limit: $\lim\limits_{x \to 6} 8 = 8$

10. Finding the limit: $\lim\limits_{x \to 4} \dfrac{x^2 - 2x - 8}{x+1} = \dfrac{16 - 8 - 8}{5} = \dfrac{0}{5} = 0$

11. Finding the limit: $\lim\limits_{x \to 3} \dfrac{x^2 - 2x - 3}{x - 3} = \lim\limits_{x \to 3} \dfrac{(x-3)(x+1)}{x-3} = \lim\limits_{x \to 3}(x+1) = 3 + 1 = 4$

12. Finding the limit: $\lim\limits_{x \to 0} \dfrac{3x^3 - 8x}{2x} = \lim\limits_{x \to 0} \dfrac{x(3x^2 - 8)}{2x} = \lim\limits_{x \to 0} \dfrac{3x^2 - 8}{2} = \dfrac{0 - 8}{2} = -4$

13. Finding the limit: $\lim\limits_{x \to \infty} \dfrac{7x^2 + 3x - 1}{8x^2 + x + 3} = \lim\limits_{x \to \infty} \dfrac{7x^2}{8x^2} = \lim\limits_{x \to \infty} \dfrac{7}{8} = \dfrac{7}{8}$

14. Finding the limit: $\lim\limits_{x \to \infty} \dfrac{x^4 + 2x^3 + x + 1}{x^5 + 3x - 9} = \lim\limits_{x \to \infty} \dfrac{x^4}{x^5} = \lim\limits_{x \to \infty} \dfrac{1}{x} = 0$

15. Finding the limit: $\lim\limits_{x \to -\infty} \dfrac{x^2 + 11x + 10}{x + 1} = \lim\limits_{x \to -\infty} \dfrac{x^2}{x} = \lim\limits_{x \to -\infty} x = -\infty$, which does not exist

16. The function is discontinuous at $x = a$, since $\lim\limits_{x \to a^-} f(x) \neq \lim\limits_{x \to a^+} f(x)$ and thus $\lim\limits_{x \to a} f(x)$ does not exist.

17. The function is discontinuous at $x = a$, since $f(a)$ does not exist. Also note that $\lim\limits_{x \to a^-} f(x) \neq \lim\limits_{x \to a^+} f(x)$ and thus $\lim\limits_{x \to a} f(x)$ does not exist.

18. Since $\lim\limits_{x \to 6} f(x) = \lim\limits_{x \to 6} \dfrac{x^2 - 5x - 24}{x + 3} = \lim\limits_{x \to 6} \dfrac{(x-8)(x+3)}{x+3} = \lim\limits_{x \to 6}(x-8) = -2 = f(6)$, the function is continuous at $x = 6$.

19. Since $\lim\limits_{x\to 1^-} f(x) = \lim\limits_{x\to 1^-}(3x-1) = 3-1 = 2$ and $\lim\limits_{x\to 1^+} f(x) = \lim\limits_{x\to 1^+}(x^2-2x+5) = 1-2+5 = 4$, $\lim\limits_{x\to 1} f(x)$ does not exist and thus the function is discontinuous at $x = 1$.

20. Finding the average rate of change: $\dfrac{f(1)-f(-1)}{1-(-1)} = \dfrac{\dfrac{1-4}{1+2}-\dfrac{1-4}{-1+2}}{2} = \dfrac{-1+3}{2} = \dfrac{2}{2} = 1$

21. Finding the difference quotient:

$$\frac{f(x+h)-f(x)}{h} = \frac{\left(-3(x+h)^2+(x+h)-2\right)-\left(-3x^2+x-2\right)}{h}$$

$$= \frac{-3x^2-6xh-3h^2+x+h-2+3x^2-x+2}{h}$$

$$= \frac{-6xh-3h^2+h}{h}$$

$$= -6x-3h+1$$

22. Finding the instantaneous rate of change:

$$\lim_{h\to 0}\frac{f(x+h)-f(x)}{h} = \lim_{h\to 0}\frac{\left(3(x+h)^2-8(x+h)+1\right)-\left(3x^2-8x+1\right)}{h}$$

$$= \lim_{h\to 0}\frac{3x^2+6xh+3h^2-8x-8h+1-3x^2+8x-1}{h}$$

$$= \lim_{h\to 0}\frac{6xh+3h^2-8h}{h}$$

$$= \lim_{h\to 0}(6x+3h-8)$$

$$= 6x-8$$

Evaluating when $x = 5$: $6(5)-8 = 30-8 = 22$

23. The slope of the tangent line is the instantaneous rate of change at the point.
Finding the instantaneous rate of change:

$$\lim_{h\to 0}\frac{f(x+h)-f(x)}{h} = \lim_{h\to 0}\frac{\dfrac{x+h+2}{x+h-6}-\dfrac{x+2}{x-6}}{h}$$

$$= \lim_{h\to 0}\frac{\dfrac{x+h+2}{x+h-6}-\dfrac{x+2}{x-6}}{h}\cdot\frac{(x+h-6)(x-6)}{(x+h-6)(x-6)}$$

$$= \lim_{h\to 0}\frac{(x+h+2)(x-6)-(x+2)(x+h-6)}{h(x+h-6)(x-6)}$$

$$= \lim_{h\to 0}\frac{\left(x^2+xh-4x-6h-12\right)-\left(x^2+xh-4x+2h-12\right)}{h(x+h-6)(x-6)}$$

$$= \lim_{h\to 0}\frac{-8h}{h(x+h-6)(x-6)}$$

$$= \lim_{h\to 0}\frac{-8}{(x+h-6)(x-6)}$$

$$= \frac{-8}{(x-6)^2}$$

Evaluating when $x = 5$: $\dfrac{-8}{(5-6)^2} = \dfrac{-8}{1} = -8$

24. **a.** Evaluating: $N(15) = 90(15)^2 - (15)^3 = 16,875$

On day 15, a total of 16,875 will have the flu.

b. Finding the average rate of change:

$$\frac{N(70) - N(40)}{70 - 40} = \frac{\left(90 \cdot 70^2 - 70^3\right) - \left(90 \cdot 40^2 - 40^3\right)}{30} = \frac{98,000 - 80,000}{30} = \frac{18,000}{30} = 600$$

The average rate of change is 600 people per day.

c. Finding the instantaneous rate of change:

$$\lim_{h \to 0} \frac{N(40 + h) - N(40)}{h} = \lim_{h \to 0} \frac{90(40 + h)^2 - (40 + h)^3 - 80,000}{h}$$

$$= \lim_{h \to 0} \frac{90\left(1,600 + 80h + h^2\right) - \left(64,000 + 4,800h + 120h^2 + h^3\right) - 80,000}{h}$$

$$= \lim_{h \to 0} \frac{\left(144,000 + 7,200h + 90h^2\right) - \left(64,000 + 4,800h + 120h^2 + h^3\right) - 80,000}{h}$$

$$= \lim_{h \to 0} \frac{2,400h - 30h^2 - h^3}{h}$$

$$= \lim_{h \to 0} \left(2,400 - 30h - h^2\right)$$

$$= 2,400$$

The instantaneous rate of change is 2,400 people per day on day 40.

25. **a.** The domain is $\{x \mid x \geq 0\}$.

b. Evaluating: $\overline{C}(280) = \dfrac{6(280) + 8,000}{2(280) + 300} = \dfrac{9,680}{860} \approx \$11.26$

c. Finding the average rate of change:

$$\frac{\overline{C}(300) - \overline{C}(280)}{300 - 280} = \frac{\dfrac{6(300) + 8,000}{2(300) + 300} - \dfrac{6(280) + 8,000}{2(280) + 300}}{20} = \frac{\dfrac{9,800}{900} - \dfrac{9,680}{860}}{20} \approx \frac{-0.3669}{20} \approx -\$0.0183$$

d. Finding the limit: $\displaystyle\lim_{x \to \infty} \frac{6x + 8,000}{2x + 300} = \lim_{x \to \infty} \frac{6x}{2x} = \$3$

In the long run, the average cost will approach $3.

e. Finding the limit: $\displaystyle\lim_{x \to -150} \frac{6x + 8,000}{2x + 300} = \frac{-900 + 8,000}{-300 + 300} = \pm\infty$

As x approaches -150, the average cost approaches ∞.

Chapter 2
Differentiation: The Language of Change

2.1 The Derivative of a Function and Two Interpretations

1. Using the limit definition:

$$f'(x) = \lim_{h \to 0} \frac{f(x+h) - f(x)}{h}$$

$$= \lim_{h \to 0} \frac{\left[(x+h)^2 + 2(x+h) - 7\right] - \left[x^2 + 2x - 7\right]}{h}$$

$$= \lim_{h \to 0} \frac{x^2 + 2xh + h^2 + 2x + 2h - 7 - x^2 - 2x + 7}{h}$$

$$= \lim_{h \to 0} \frac{2xh + h^2 + 2h}{h}$$

$$= \lim_{h \to 0} \frac{h(2x + h + 2)}{h}$$

$$= \lim_{h \to 0} (2x + h + 2)$$

$$= 2x + 2$$

3. Using the limit definition:

$$f'(x) = \lim_{h \to 0} \frac{f(x+h) - f(x)}{h}$$

$$= \lim_{h \to 0} \frac{\left[8(x+h) + 9\right] - \left[8x + 9\right]}{h}$$

$$= \lim_{h \to 0} \frac{8x + 8h + 9 - 8x - 9}{h}$$

$$= \lim_{h \to 0} \frac{8h}{h}$$

$$= 8$$

5. Using the limit definition:

$$f'(x) = \lim_{h \to 0} \frac{f(x+h) - f(x)}{h}$$

$$= \lim_{h \to 0} \frac{\dfrac{3}{x+h} - \dfrac{3}{x}}{h}$$

$$= \lim_{h \to 0} \frac{\dfrac{3}{x+h} - \dfrac{3}{x}}{h} \cdot \frac{x(x+h)}{x(x+h)}$$

$$= \lim_{h \to 0} \frac{3x - 3(x+h)}{hx(x+h)}$$

$$= \lim_{h \to 0} \frac{3x - 3x - 3h}{hx(x+h)}$$

$$= \lim_{h \to 0} \frac{-3h}{hx(x+h)}$$

$$= \lim_{h \to 0} \frac{-3}{x(x+h)}$$

$$= -\frac{3}{x^2}$$

7. Using the differentiation rules: $f'(x) = 10(3x^2) - 12 + 0 = 30x^2 - 12$

9. Using the differentiation rules: $f'(x) = 8\left(\dfrac{5}{2}x^{3/2}\right) = 20x^{3/2}$

11. Using the differentiation rules: $\dfrac{dy}{dx} = 6\left(-\dfrac{3}{2}x^{-5/2}\right) + 4\left(\dfrac{1}{2}x^{-1/2}\right) + 8\left(\dfrac{3}{2}x^{1/2}\right) = -9x^{-5/2} + 2x^{-1/2} + 12x^{1/2}$

13. Since $y = \dfrac{1}{x} = x^{-1}$, using the differentiation rules: $\dfrac{dy}{dx} = -1x^{-2} = -\dfrac{1}{x^2}$

15. Since $f(x) = \dfrac{2}{x^3} + \dfrac{3}{x^2} + \dfrac{3}{x} - \dfrac{1}{4} = 2x^{-3} + 3x^{-2} + 3x^{-1} - \dfrac{1}{4}$, using the differentiation rules:

    $$f'(x) = 2(-3x^{-4}) + 3(-2x^{-3}) + 3(-1x^{-2}) - 0 = -\dfrac{6}{x^4} - \dfrac{6}{x^3} - \dfrac{3}{x^2}$$

17. Since $f(x) = \dfrac{-2}{\sqrt[5]{x^2}} + \dfrac{5x^{4/5}}{\sqrt[5]{x^3}} = -2x^{-2/5} + 5x^{4/5 - 3/5} = -2x^{-2/5} + 5x^{1/5}$, using the differentiation rules:

    $$f'(x) = -2\left(-\dfrac{2}{5}x^{-7/5}\right) + 5\left(\dfrac{1}{5}x^{-4/5}\right) = \dfrac{4}{5x^{7/5}} + \dfrac{1}{x^{4/5}} = \dfrac{4}{5x\sqrt[5]{x^2}} + \dfrac{1}{\sqrt[5]{x^4}}$$

19. First note that $f'(x) = 2x + 6$. Thus the values are:

    $$f(3) = 3^2 + 6(3) - 4 = 9 + 18 - 4 = 23$$
    $$f'(3) = 2(3) + 6 = 6 + 6 = 12$$
    $$f(4) - f(3) = \left[4^2 + 6(4) - 4\right] - \left[3^2 + 6(3) - 4\right] = 36 - 23 = 13$$

21. Evaluating the derivative:

    $$\dfrac{dy}{dx} = 2(4x^3) + 3(3x^2) + 1 + 0 = 8x^3 + 9x^3 + 1$$
    $$\left.\dfrac{dy}{dx}\right|_{x=0} = 8(0)^3 + 9(0)^3 + 1 = 1$$

23. First find $f'(x) = 2x + 6$. Now set this equal to 0:

    $$2x + 6 = 0$$
    $$2x = -6$$
    $$x = -3$$

25. **a.** Evaluating: $f(1) = 4(1)^2 - 16(1) + 17 = 5$

    b. Since $f'(x) = 8x - 16$, evaluating: $f'(1) = 8(1) - 16 = -8$

    c. Setting $f'(x) = 0$:

    $$f'(x) = 0$$
    $$8x - 16 = 0$$
    $$8x = 16$$
    $$x = 2$$

    d. The slope of the tangent line is $f'(3) = 8(3) - 16 = 8$. Using the point $(3, f(3)) = (3, 5)$ in the point-slope formula:

    $$y - 5 = 8(x - 3)$$
    $$y - 5 = 8x - 24$$
    $$y = 8x - 19$$

27. The expression $P(60) = 40$ means a profit of $40,000 will be earned if $60,000 is spent on advertising. The expression $\dfrac{dP}{dx}\bigg|_{x=60} = 2$ means the profit will increase by $2,000 if an additional $1,000 is spent on advertising at the current advertising level of $60,000.

29. The expression $N(150) = 1,200$ means 150 hours are necessary to build 1,200 objects. The expression $N'(150) = 21.25$ means one additional hour will result in building 21.25 additional objects if 150 hours have already been spent.

31. The expression $S(44) = 87$ means an average score of 87 points will be attained if 44 teachers have college degrees in the subject they teach. The expression $S'(44) = 1.5$ means the average score will increase 1.5 points if an additional teacher is hired with a degree in the subject they teach.

33. The expression $f(2) = 7$ means the cloud will travel 7 miles 2 hours after it is released. The expression $f'(2) = 3$ means the cloud will travel an additional 3 miles during the next (third) hour.

35. a. Evaluating the function: $R(4) = -(4)^4 + 11(4)^3 - 39(4)^2 + 45(4) = 4$
   The revenue was $4 million at the beginning of 2011.

   b. Finding the derivative: $R'(x) = -4x^3 + 33x^2 - 78x + 45$

   Evaluating the derivative: $R'(4) = -4(4)^3 + 33(4)^2 - 78(4) + 45 = 5$
   The revenue was increasing by $5 million per year in the year 2011.

37. a. Evaluating the function: $A(54) = 0.11(54)^{2/3} \approx 1.572$
   The surface area of a 54 kg person is approximately 1.572 square meters.

   b. Finding the derivative: $A'(m) = \dfrac{0.22}{3}m^{-1/3} = \dfrac{0.22}{3m^{1/3}}$

   Evaluating the derivative: $A'(54) = \dfrac{0.22}{3(54)^{1/3}} \approx 0.019$

   The surface area of a 54 kg person will increase approximately 0.019 square meters if the person gains 1 kg.

39. a. Evaluating the function: $h(16) = \dfrac{250}{(16)^{1/4}} = \dfrac{250}{2} = 125$

   The heart rate is 125 bpm for a 16-pound baby.

   b. Evaluating the function: $h(256) = \dfrac{250}{(256)^{1/4}} = \dfrac{250}{4} = 62.5$

   The heart rate is 62.5 bpm for a 256-pound man.

   c. Finding the derivative for $h(w) = \dfrac{250}{w^{1/4}} = 250w^{-1/4}$: $h'(w) = 250\left(-\dfrac{1}{4}w^{-5/4}\right) = -\dfrac{62.5}{w^{5/4}}$

   Evaluating the derivative: $h'(81) = -\dfrac{62.5}{(81)^{5/4}} = -\dfrac{62.5}{243} \approx -0.26$

   An 81-pound child will have a heart rate decrease of 0.26 bpm when gaining the next pound.

41. a. Finding the derivative: $R'(x) = 0 + 7.130\left(\dfrac{1}{2}x^{-1/2}\right) = \dfrac{3.565}{\sqrt{x}}$

   Evaluating the derivative: $R'(6,000) = \dfrac{3.565}{\sqrt{6,000}} \approx 0.0460$

   The monthly revenue should increase by approximately $46 if the advertising budget is increased from $6,000 to $6,001.

   b. Evaluating the revenue: $R(6,000) = 85 + 7.130\sqrt{6,000} \approx 637.287$
   The revenue is approximately $637,287 when the advertising budget is $6,000.

43. Finding the derivative for $A(t) = 3.1 + \dfrac{4.8}{\sqrt{t}} = 3.1 + 4.8t^{-1/2}$: $A'(t) = 0 + 4.8\left(-\dfrac{1}{2}t^{-3/2}\right) = -\dfrac{2.4}{t^{3/2}}$

Evaluating the derivative: $A'(2) = -\dfrac{2.4}{2^{3/2}} \approx -0.849$

The glucose concentration is decreasing by approximately 0.849 mg per hour 2 hours after ingestion of the glucose.

45. Finding the derivative: $f'(x) = \dfrac{3}{5}x^{-2/5} + \dfrac{2}{5}x^{-3/5} - x^{-4/5}$

Evaluating the derivative: $f'(1.5) = \dfrac{3}{5}(1.5)^{-2/5} + \dfrac{2}{5}(1.5)^{-3/5} - (1.5)^{-4/5} \approx 0.1008$

47. The calculator response is undefined, or error.

Finding the derivative of $f(x) = 3x^{3/4} + x^{2/3}$: $f'(x) = \dfrac{9}{4}x^{-1/4} + \dfrac{2}{3}x^{-1/3} = \dfrac{9}{4x^{1/4}} + \dfrac{2}{3x^{1/3}}$

Note that $f'(0)$ is undefined, since 0 is not in the domain of $f'(x) = \dfrac{9}{4x^{1/4}} + \dfrac{2}{3x^{1/3}}$.

49. Multiplying: $(5x^2 + 4)(x^3 + 11) = 5x^5 + 4x^3 + 55x^2 + 44$

51. Multiplying: $(x^2 - 1)(3x^2 + 2) = 3x^4 + 2x^2 - 3x^2 - 2 = 3x^4 - x^2 - 2$

53. Simplifying: $2t(t^2 - 4) + 2t(t^2 + 1) = 2t^3 - 8t + 2t^3 + 2t = 4t^3 - 6t$

55. Solving the equation:
$$(t^2 - 4)(t^2 + 1) = 0$$
$$(t + 2)(t - 2)(t^2 + 1) = 0$$
$$t = -2, 2$$

57. Evaluating the function: $g(2) = \dfrac{(2)^4 - 5(2)^2 + 2(2) - 2}{(2^2 - 1)^2} = \dfrac{16 - 20 + 4 - 2}{(4 - 1)^2} = -\dfrac{2}{9}$

59. Evaluating the function: $g(4) = \dfrac{-1}{(4 - 3)^2} = \dfrac{-1}{1} = -1$

61. Using the point-slope formula:
$$y - 3 = -1(x - 4)$$
$$y - 3 = -x + 4$$
$$y = -x + 7$$

63. Finding the derivative: $\dfrac{dy}{dx} = 5(2x) + 0 = 10x$ 65. Finding the derivative: $\dfrac{dy}{dx} = 2$

67. Finding the derivative: $\dfrac{dy}{dt} = 2t + 0 = 2t$

69. Finding the derivative: $\dfrac{dy}{dx} = 5(5x^4) + 4(3x^2) + 55(2x) + 0 = 25x^4 + 12x^2 + 110x$

2.2 Differentiating Products and Quotients

1. Using the product rule: $f'(x) = (2)(4x-5) + (2x+1)(4) = 8x - 10 + 8x + 4 = 16x - 6$

3. Using the product rule: $f'(x) = (2x)(x-2) + (x^2+4)(1) = 2x^2 - 4x + x^2 + 4 = 3x^2 - 4x + 4$

5. Using the product rule: $f'(x) = (2x)(x^2+2) + (x^2-1)(2x) = 2x^3 + 4x + 2x^3 - 2x = 4x^3 + 2x$

7. Using the quotient rule: $f'(x) = \dfrac{(5x+2)(4)-(4x+3)(5)}{(5x+2)^2} = \dfrac{20x+8-20x-15}{(5x+2)^2} = \dfrac{-7}{(5x+2)^2}$

9. Using the quotient rule: $f'(x) = \dfrac{(3x+2)(1)-(x-4)(3)}{(3x+2)^2} = \dfrac{3x+2-3x+12}{(3x+2)^2} = \dfrac{14}{(3x+2)^2}$

11. Using the quotient rule:

$$f'(x) = \frac{(x^2-4)(9x^2-7)-(3x^3-7x+1)(2x)}{(x^2-4)^2} = \frac{9x^4-43x^2+28-6x^4+14x^2-2x}{(x^2-4)^2} = \frac{3x^4-29x^2-2x+28}{(x^2-4)^2}$$

13. Using the quotient rule: $f'(x) = \dfrac{(x^3+1)(4x^3)-(x^4-6)(3x^2)}{(x^3+1)^2} = \dfrac{4x^6+4x^3-3x^6+18x^2}{(x^3+1)^2} = \dfrac{x^6+4x^3+18x^2}{(x^3+1)^2}$

15. Using the quotient rule (and the product rule):

$$f'(x) = \frac{(7x+2)\left[(2)(x^2+5)+(2x-1)(2x)\right]-(2x-1)(x^2+5)(7)}{(7x+2)^2}$$

$$= \frac{(7x+2)\left[2x^2+10+4x^2-2x\right]-(2x-1)(7x^2+35)}{(7x+2)^2}$$

$$= \frac{(7x+2)(6x^2-2x+10)-(2x-1)(7x^2+35)}{(7x+2)^2}$$

$$= \frac{(42x^3-2x^2+66x+20)-(14x^3-7x^2+70x-35)}{(7x+2)^2}$$

$$= \frac{28x^3+5x^2-4x+55}{(7x+2)^2}$$

17. Using the quotient rule: $f'(x) = \dfrac{(x^2+3x+4)(0)-(8)(2x+3)}{(x^2+3x+4)^2} = \dfrac{-16x-24}{(x^2+3x+4)^2}$

19. Using the quotient rule:

$$f'(x) = \frac{(x-4)\left(\frac{1}{2}x^{-1/2}\right)-(x^{1/2})(1)}{(x-4)^2} = \frac{(x-4)\left(\frac{1}{2}x^{-1/2}\right)-(x^{1/2})(1)}{(x-4)^2} \cdot \frac{2x^{1/2}}{2x^{1/2}} = \frac{(x-4)-(2x)}{2\sqrt{x}(x-4)^2} = \frac{-x-4}{2x^{1/2}(x-4)^2}$$

21. Finding the derivative: $f'(x) = (9)(x-6)+(9x+2)(1) = 9x - 54 + 9x + 2 = 18x - 52$

Evaluating the derivative: $f'(4) = 18(4) - 52 = 72 - 52 = 20$

The tangent line slope at $x = 4$ is 20. The function is increasing 20 units for each unit increase in x.

23. Finding the derivative: $g'(t) = \dfrac{(8t+6)(5)-(5t+4)(8)}{(8t+6)^2} = \dfrac{40t+30-40t-32}{(8t+6)^2} = \dfrac{-2}{(8t+6)^2}$

Evaluating the derivative: $g'(10) = \dfrac{-2}{(80+6)^2} = \dfrac{-1}{3,698}$

The tangent line slope at $t = 10$ is $-\dfrac{1}{3,698}$. The function is decreasing $\dfrac{1}{3,698}$ units for each unit increase in t.

25. Finding the derivative: $f'(x) = (12x^3)(x^3+1) + (3x^4)(3x^2)$

27. Finding the derivative: $f'(x) = (3)(\sqrt{5x+2}) + (3x+7)\left(\dfrac{5}{2\sqrt{5x+2}}\right)$

29. Finding the derivative: $f'(x) = \dfrac{(8x+1)(5) - (5x-7)(8)}{(8x+1)^2}$

31. Finding the derivative: $f'(x) = \dfrac{1}{4}(x^2+3)^{-3/4}(2x)(2x+7) + (x^2+3)^{1/4}(2)$

33. Finding the derivative: $N'(t) = \dfrac{(105t-80)(80) - (80t)(105)}{(105t-80)^2} = \dfrac{8400t - 6400 - 8400t}{25(21t-16)^2} = \dfrac{-256}{(21t-16)^2}$

Evaluating: $N(1) = \dfrac{80(1)}{105(1)-80} = \dfrac{80}{25} = 3.2$

A person can remember 3.2 facts 1 hour after memorizing them.

Evaluating: $N(3) = \dfrac{80(3)}{105(3)-80} = \dfrac{240}{235} \approx 1.02$

A person can remember approximately 1.02 facts 3 hours after memorizing them.

Evaluating: $N'(1) = \dfrac{-256}{(21-16)^2} = -10.24$

The rate that a person remembers facts is decreasing by 10.24 facts per hour 1 hour after memorizing them.

Evaluating: $N'(3) = \dfrac{-256}{(63-16)^2} \approx -0.12$

The rate that a person remembers facts is decreasing by approximately 0.12 facts per hour 3 hours after memorizing them.

35. **a.** Evaluating: $V(4) = \dfrac{340(4)}{4+3} = \dfrac{1360}{7} \approx 194.3$

The speed of the dragster is approximately 194.3 mph 4 seconds into its run.

  b. Finding the derivative: $V'(t) = \dfrac{(t+3)(340) - (340t)(1)}{(t+3)^2} = \dfrac{340t + 1{,}020 - 340t}{(t+3)^2} = \dfrac{1{,}020}{(t+3)^2}$

  c. Evaluating: $V'(4) = \dfrac{1{,}020}{(4+3)^2} \approx 20.8$

The speed is changing by 20.8 mph per second 4 seconds into its run.

  d. The dragster's speed is increasing by 20.8 mph per second after 4 seconds into its run.

37. Finding the derivative: $C'(t) = \dfrac{(t^2+t+4)(21) - (21t)(2t+1)}{(t^2+t+4)^2} = \dfrac{21t^2 + 21t + 84 - 42t^2 - 21t}{(t^2+t+4)^2} = \dfrac{-21t^2 + 84}{(t^2+t+4)^2}$

Evaluating the derivative: $C'(3) = \dfrac{-21(3)^2 + 84}{(3^2+3+4)^2} = \dfrac{-189 + 84}{(16)^2} \approx -0.410$

The concentration will decrease by approximately 0.41 mg/mm³ between the third and fourth hours after ingestion.

39. **a.** Evaluating the functions: $\dfrac{F(4)}{A(4)} = \dfrac{-3(4)^2 + 15(4)}{-4^2 + 9(4)} = \dfrac{12}{20} = 0.6$

The revenue of the financial advisors network is 0.6 that of the accountants niche in 2011.

b. Finding the derivative:

$$\left(\frac{F(x)}{A(x)}\right)' = \frac{\left(-x^2 + 9x\right)(-6x + 15) - \left(-3x^2 + 15x\right)(-2x + 9)}{\left(-x^2 + 9x\right)^2}$$

$$= \frac{\left(6x^3 - 69x^2 + 135x\right) - \left(6x^3 - 57x^2 + 135x\right)}{\left(-x^2 + 9x\right)^2}$$

$$= \frac{-12x^2}{\left(-x^2 + 9x\right)^2}$$

$$= \frac{-12x^2}{x^2(x-9)^2}$$

$$= \frac{-12}{(x-9)^2}$$

c. Evaluating the derivative: $\left(\dfrac{F(x)}{A(x)}\right)'(4) = \dfrac{-12}{(4-9)^2} = -0.48$

d. The ratio of financial advisors network revenue to accountants network revenue is decreasing by 0.48 million dollars in the year 2011.

41. **a.** The marginal revenue is: $R'(x) = 95 - 0.16x$

The marginal cost is: $C'(x) = 35 - 0.08x$

The profit function is given by: $P(x) = \left(95x - 0.08x^2\right) - \left(1,200 + 35x - 0.04x^2\right) = -0.04x^2 + 60x - 1,200$

The marginal profit is: $P'(x) = 60 - 0.08x$

b. Evaluating the marginal revenue: $R'(601) = 95 - 0.16(601) = -1.16$

Evaluating the marginal revenue: $R'(900) = 95 - 0.16(900) = -49$

The marginal revenue is decreasing as the production level increases.

c. Evaluating the marginal cost: $C'(601) = 35 - 0.08(601) = -13.08$

Evaluating the marginal cost: $C'(900) = 35 - 0.08(900) = -37$

The marginal cost is decreasing as the production level increases.

d. Evaluating the marginal profit: $P'(601) = 60 - 0.08(601) = 11.92$

Evaluating the marginal profit: $P'(900) = 60 - 0.08(900) = -12$

The marginal profit is decreasing as the production level increases.

43. **a.** Finding the marginal profit:

$$P'(t) = \frac{(t^2+10)(500-20t)-(200+500t-10t^2)(2t)}{(t^2+10)^2}$$

$$= \frac{(-20t^3+500t^2-200t+5,000)-(-20t^3+1,000t^2+400t)}{(t^2+10)^2}$$

$$= \frac{-500t^2-600t+5,000}{(t^2+10)^2}$$

$$= \frac{-100(5t^2+6t-50)}{(t^2+10)^2}$$

b. Evaluating the marginal profit: $P'(5) = \dfrac{-100(5(5)^2+6(5)-50)}{(5^2+10)^2} = \dfrac{-10,500}{1,225} \approx -8.571$

The marginal profit is –\$8,571 after 5 months. The publisher can expect the profit to decrease by \$8,571 from month 5 to month 6.

c. Evaluating the marginal profit: $P'(10) = \dfrac{-100(5(10)^2+6(10)-50)}{(10^2+10)^2} = \dfrac{-51,000}{12,100} \approx -4.215$

The marginal profit is –\$4,215 after 10 months. The publisher can expect the profit to decrease by \$4,215 from month 10 to month 11.

d. When $t > 3$, the value of $P'(t)$ is negative. Thus the monthly profit decreases for $t > 3$.

45. **a.** Analyst Star's revenue projection is 1.04 times that of analyst Moon's revenue projection.
   b. Analyst Star's marginal revenue projection is 3.12 times that of analyst Moon's marginal revenue projection.
   c. The ratio of analyst Star's revenue to analyst Moon's revenue is increasing by 0.02 after 2 years. Thus, in year 2, we expect analyst Star's marginal revenue prediction to be $1.04 + 0.02 = 1.06$ times analyst Moon's marginal revenue prediction.

47. Finding the derivative:

$$f'(x) = \frac{(x^2+6x)(2x-5)-(x^2-5x)(2x+6)}{(x^2+6x)^2} = \frac{(2x^3+7x^2-30x)-(2x^3-4x^2-30x)}{x^2(x+6)^2} = \frac{11x^2}{x^2(x+6)^2} = \frac{11}{(x+6)^2}$$

49. Finding the derivative:

$$f'(x) = \frac{(x^{2/3}-x)\left(\frac{2}{3}x^{-1/3}+1\right)-(x^{2/3}+x)\left(\frac{2}{3}x^{-1/3}-1\right)}{(x^{2/3}-x)^2}$$

$$= \frac{(x^{2/3}-x)\left(\frac{2}{3}x^{-1/3}+1\right)-(x^{2/3}+x)\left(\frac{2}{3}x^{-1/3}-1\right)}{(x^{2/3}-x)^2} \cdot \frac{3x^{1/3}}{3x^{1/3}}$$

$$= \frac{(x^{2/3}-x)(2+3x^{1/3})-(x^{2/3}+x)(2-3x^{1/3})}{3x^{1/3}(x^{2/3}-x)^2}$$

$$= \frac{(2x^{2/3}+x-3x^{4/3})-(2x^{2/3}-x-3x^{4/3})}{3x^{1/3}(x^{2/3}-x)^2}$$

$$= \frac{2x}{3x^{1/3}(x^{2/3}-x)^2}$$

Evaluating the derivative: $f'(64) = \dfrac{2(64)}{3(64)^{1/3}(64^{2/3}-64)^2} = \dfrac{128}{12(-48)^2} = \dfrac{1}{216}$

51. Finding the derivative: $P'(x) = \dfrac{(x+5)(0)-(1.5)(1)}{(x+5)^2} = \dfrac{-1.5}{(x+5)^2}$

Evaluating: $P(3) = \dfrac{1.5}{3+5} = 0.1875$

The concentration is 18.75% when 3 liters of distilled water are added.

Evaluating: $P'(3) = \dfrac{-1.5}{(3+5)^2} \approx -0.0234$

The concentration is decreasing by 2.34% after 3 liters of distilled water have been added.

53. Finding the value: $f(2) = \dfrac{1}{2}(2)^4 - 4(2)^2 = 8 - 16 = -8$

55. Solving the equation:
$$x^3 - 9x = 0$$
$$x(x^2 - 9) = 0$$
$$x(x+3)(x-3) = 0$$
$$x = 0, -3, 3$$

57. Finding the derivative: $g'(x) = 2(3x^2) - 8 = 6x^2 - 8$

59. Finding the derivative: $f'(x) = 3x^2 - 9$

2.3 Higher-Order Derivatives

1. Finding the first four derivatives:
$$f'(x) = 6x^5 - 10x^4 + 6$$
$$f''(x) = 30x^4 - 40x^3$$
$$f'''(x) = 120x^3 - 120x^2$$
$$f^{(4)}(x) = 360x^2 - 240x$$

3. Finding the first four derivatives:
$$f'(x) = 4x^3 + 6x^2 - 11x^{-2}$$
$$f''(x) = 12x^2 + 12x + 22x^{-3}$$
$$f'''(x) = 24x + 12 - 66x^{-4}$$
$$f^{(4)}(x) = 24 + 264x^{-5}$$

5. Finding the first four derivatives of $f(x) = x^3 + 6x^2 - 4x^{1/2}$:

$$f'(x) = 3x^2 + 12x - 4\left(\frac{1}{2}x^{-1/2}\right) = 3x^2 + 12x - 2x^{-1/2}$$

$$f''(x) = 6x + 12 - 2\left(-\frac{1}{2}x^{-3/2}\right) = 6x + 12 + x^{-3/2}$$

$$f'''(x) = 6 + \left(-\frac{3}{2}x^{-5/2}\right) = 6 - \frac{3}{2}x^{-5/2}$$

$$f^{(4)}(x) = -\frac{3}{2}\left(-\frac{5}{2}x^{-7/2}\right) = \frac{15}{4}x^{-7/2}$$

7. Finding $f'(x)$ and $f''(x)$ for $f(x) = \dfrac{x-4}{x} = 1 - 4x^{-1}$:

$$f'(x) = -4(-x^{-2}) = 4x^{-2} = \frac{4}{x^2} \qquad f''(x) = 4(-2x^{-3}) = -8x^{-3} = \frac{-8}{x^3}$$

9. Finding $f'(x)$ and $f''(x)$ for $f(x) = \dfrac{2x+7}{x^3} = 2x^{-2} + 7x^{-3}$:

$$f'(x) = 2\left(-2x^{-3}\right) + 7\left(-3x^{-4}\right) = -4x^{-3} - 21x^{-4} = \frac{-4}{x^3} - \frac{21}{x^4} = \frac{-4x - 21}{x^4}$$

$$f''(x) = -4\left(-3x^{-4}\right) - 21\left(-4x^{-5}\right) = 12x^{-4} + 84x^{-5} = \frac{12}{x^4} + \frac{84}{x^5} = \frac{12x + 84}{x^5}$$

11. Finding $f'(1)$ and $f''(1)$ for $f(x) = x^2 + 5x - 4$:

$$f'(x) = 2x + 5 \qquad\qquad\qquad f''(x) = 2$$
$$f'(1) = 2(1) + 5 = 7 \qquad\qquad f''(1) = 2$$

$f'(1)$ represents the rate $f(x)$ is changing when $x = 1$, and $f''(1)$ represents the rate $f'(x)$ is changing when $x = 1$. At $x = 1$ the function is increasing at an increasing rate.

13. Let $f(x) = (4x-1)(2x-3) = 8x^2 - 14x + 3$:

a. Finding $f(1)$: $f(1) = 8(1)^2 - 14(1) + 3 = 8 - 14 + 3 = -3$

b. Finding $f'(x)$: $f'(x) = 16x - 14$ Finding $f'(1)$: $f'(x) = 16(1) - 14 = 2$

c. Finding $f''(x)$: $f''(x) = 16$ Finding $f''(1)$: $f''(1) = 16$

d. Setting $f(x) = 0$:   **e.** Setting $f'(x) = 0$:

$$(4x-1)(2x-3) = 0 \qquad\qquad 16x - 14 = 0$$
$$x = \frac{1}{4}, \frac{3}{2} \qquad\qquad\qquad 16x = 14$$
$$x = \frac{7}{8}$$

15. The first derivative tells us the marginal cost is $0.65 when the production level is 4,500 pillows. So the expected cost of producing the next pillow is $0.65. The second derivative tells us the marginal cost is increasing, so the expected cost of producing additional pillows will increase above the $0.65.

17. The first derivative tells us that 685,000 less gallons will flow out of the reservoir after 230 days. The second derivative tells us the rate of change is decreasing, so the decrease will less than 685,000 gallons.

19. The first derivative tells us that the useful life will increase by 30 hours when the complexity index is 7 units. The second derivative tells us this rate will decrease as the complexity index increases.

21. The first derivative tells us that there will be 15 fewer alcohol-related traffic accidents 15 months after the law has taken effect. The second derivative tells us this amount will increase (greater than 15) in subsequent months.

23. Finding the first and second derivatives:

$$G'(t) = 2.4t + 64.3 \qquad\qquad G''(t) = 2.4$$

Evaluating the derivatives:

$$G'(32) = 2.4(32) + 64.3 = 141.1 \qquad\qquad G''(32) = 2.4$$

This tells us China's global manufacturing value increased by $141.1 billion in the year 2012, and the second derivative tells us this increase will increase another $2.4 billion (to $143.5 billion) the subsequent year.

25. Evaluating each derivative:

$$S'(40) = \frac{520 - 3(40)}{4\sqrt{260 - 40}} \approx 6.742 \qquad\qquad S''(40) = \frac{3(40) - 1{,}040}{8\sqrt{(260 - 40)^3}} \approx -0.035$$

The strength of the reaction is increasing when $x = 40$, but the reaction strength rate is decreasing.

27. Evaluating each derivative:

$$P'(4) = \frac{150 - 30(4)^2}{\left(4^2 + 5\right)^2} \approx -0.748 \qquad\qquad P''(4) = \frac{240\left(4^2 - 15\right)}{\left(4^2 + 5\right)^3} \approx 0.026$$

The population of bacteria is decreasing 4 hours after the toxin is introduced, but the rate of decrease is decreasing.

29. Finding each derivative:

$$N'(p) = -2,200p^{-3} = -\frac{2,200}{p^3} \qquad\qquad N''(p) = 6,600p^{-4} = \frac{6,600}{p^4}$$

Evaluating each derivative:

$$N'(12) = -\frac{2,200}{12^3} \approx -1.273 \qquad\qquad N''(12) = \frac{6,600}{12^4} \approx 0.318$$

The number of hair dryers sold is decreasing when the price is $12, but this rate is decreasing.

31. Finding each derivative:

$$R'(x) = -4x^3 + 33x^2 - 78x + 45 \qquad\qquad R''(x) = -12x^2 + 66x - 78$$

Evaluating each derivative when $x = 5$:

$$R'(5) = -4(5)^3 + 33(5)^2 - 78(5) + 45 = -20 \qquad R''(5) = -12(5)^2 + 66(5) - 78 = -48$$

In 2012 the revenue is decreasing and the rate of decrease is increasing.

33. Using the quotient rule:

$$f'(x) = \frac{(x^2 + 6x)(2x - 5) - (x^2 - 5x)(2x + 6)}{(x^2 + 6x)^2} = \frac{(2x^3 + 7x^2 - 30x) - (2x^3 - 4x^2 - 30x)}{x^2(x + 6)^2} = \frac{11x^2}{x^2(x + 6)^2} = \frac{11}{(x + 6)^2}$$

Using the quotient rule for $f(x) = \frac{11}{(x + 6)^2} = \frac{11}{x^2 + 12x + 36}$:

$$f''(x) = \frac{(x^2 + 12x + 36)(0) - 11(2x + 12)}{(x + 6)^4} = \frac{-22(x + 6)}{(x + 6)^4} = -\frac{22}{(x + 6)^3}$$

35. Finding the first derivative:

$$f'(x) = \frac{\left(x^{1/3} - x\right)\left(\frac{1}{3}x^{-2/3} + 1\right) - \left(x^{1/3} + x\right)\left(\frac{1}{3}x^{-2/3} - 1\right)}{\left(x^{1/3} - x\right)^2}$$

$$= \frac{\left(\frac{1}{3}x^{-1/3} - \frac{1}{3}x^{1/3} + x^{1/3} - x\right) - \left(\frac{1}{3}x^{-1/3} + \frac{1}{3}x^{1/3} - x^{1/3} - x\right)}{\left(x^{1/3} - x\right)^2}$$

$$= \frac{-\frac{2}{3}x^{1/3} + 2x^{1/3}}{\left(x^{1/3} - x\right)^2}$$

$$= \frac{4x^{1/3}}{3\left(x^{1/3} - x\right)^2}$$

$$= \frac{4x^{1/3}}{3\left(x^{2/3} - 2x^{4/3} + x^2\right)}$$

Evaluating: $f'(64) = \frac{4(64)^{1/3}}{3\left(64^{1/3} - 64\right)^2} = \frac{16}{3(-60)^2} \approx 0.00148148$

Finding the second derivative:

$$f''(x) = \frac{3\left(x^{1/3} - x\right)^2 \left(\frac{4}{3}x^{-2/3}\right) - 4x^{1/3} \cdot 3\left(\frac{2}{3}x^{-1/3} - \frac{8}{3}x^{1/3} + 2x\right)}{9\left(x^{1/3} - x\right)^4}$$

$$= \frac{4x^{-2/3}\left(x^{2/3} - 2x^{4/3} + x^2\right) - 8\left(1 - 4x^{2/3} + 3x^{4/3}\right)}{9\left(x^{1/3} - x\right)^4}$$

$$= \frac{4 - 8x^{2/3} + 4x^{4/3} - 8 + 32x^{2/3} - 24x^{4/3}}{9\left(x^{1/3} - x\right)^4}$$

$$= \frac{-4 + 24x^{2/3} - 20x^{4/3}}{9\left(x^{1/3} - x\right)^4}$$

$$= \frac{-4\left(1 - 6x^{2/3} + 5x^{4/3}\right)}{9\left(x^{1/3} - x\right)^4}$$

Evaluating: $f''(64) = \dfrac{-4\left(1 - 6(64)^{2/3} + 5(64)^{4/3}\right)}{9\left(64^{1/3} - 64\right)^4} = \dfrac{-4\left(1 - 6(16) + 5(256)\right)}{9\left(4 - 64\right)^4} \approx -0.00004064$

The function is increasing when $x = 64$ but the rate of change is decreasing.

37. Simplifying: $4(2x - 9) + 6 = 8x - 36 + 6 = 8x - 30$

39. Writing as an exponent: $\sqrt[5]{(7x - 8)^3} = (7x - 8)^{3/5}$

41. Calculating: $17^{1/3} \approx 2.57$

43. Calculating: $3{,}363^{3/4} \approx 441.62$

45. Differentiating: $\dfrac{dy}{dx} = 2$

47. Differentiating: $\dfrac{dy}{dx} = 7$

49. Differentiating: $\dfrac{dy}{dx} = 4\left(3x^2\right) + 5 = 12x^2 + 5$

51. Differentiating: $\dfrac{dy}{dx} = 6x^5$

53. Differentiating: $\dfrac{dy}{dx} = \dfrac{3}{5}x^{-2/5} = \dfrac{3}{5x^{2/5}}$

2.4 The Chain Rule and General Power Rule

1. Using the chain rule: $\dfrac{dy}{dx} = \dfrac{dy}{du} \cdot \dfrac{du}{dx} = 12(3) = 36$

3. Using the chain rule: $\dfrac{dy}{dx} = \dfrac{dy}{du} \cdot \dfrac{du}{dx} = (10)(-2x) = -20x$

5. Using the chain rule: $\dfrac{dy}{dx} = \dfrac{dy}{du} \cdot \dfrac{du}{dx} = (6u)(10x) = 6\left(5x^2 + 4\right) \cdot 10x = 60x\left(5x^2 + 4\right) = 300x^3 + 240x$

7. Using the chain rule:

$$\frac{dy}{dx} = \frac{dy}{du} \cdot \frac{du}{dx}$$

$$= \left(3u^2 - 6\right)(10x)$$

$$= 10x\left[3\left(5x^2 - 2\right)^2 - 6\right]$$

$$= 10x\left[3\left(25x^4 - 20x^2 + 4\right) - 6\right]$$

$$= 10x\left(75x^4 - 60x^2 + 6\right)$$

$$= 750x^5 - 600x^3 + 60x$$

9. Using the chain rule: $\dfrac{dy}{dx} = 3(5x^2 + 2x - 6)^2 (10x + 2) = 6(5x + 1)(5x^2 + 2x - 6)^2$

11. First write $y = \sqrt[5]{(x^2 + 7)^3} = (x^2 + 7)^{3/5}$. Using the chain rule: $\dfrac{dy}{dx} = \dfrac{3}{5}(x^2 + 7)^{-2/5}(2x) = \dfrac{6x}{5(x^2 + 7)^{2/5}}$

13. Using the product rule:

$$\dfrac{dy}{dx} = (x^2 + 3)^3 \cdot 5(2x - 1)^4(2) + 3(x^2 + 3)^2 (2x) \cdot (2x - 1)^5$$

$$= 10(x^2 + 3)^3 (2x - 1)^4 + 6x(x^2 + 3)^2 (2x - 1)^5$$

$$= 2(x^2 + 3)^2 (2x - 1)^4 [5(x^2 + 3) + 3x(2x - 1)]$$

$$= 2(x^2 + 3)^2 (2x - 1)^4 (5x^2 + 15 + 6x^2 - 3x)$$

$$= 2(x^2 + 3)^2 (2x - 1)^4 (11x^2 - 3x + 15)$$

15. Using the product rule:

$$\dfrac{dy}{dx} = x^4 \cdot \dfrac{1}{5}(5x + 2)^{-4/5}(5) + 4x^3 \cdot (5x + 2)^{1/5}$$

$$= x^4 (5x + 2)^{-4/5} + 4x^3 \cdot (5x + 2)^{1/5}$$

$$= x^3 (5x + 2)^{-4/5} [x + 4(5x + 2)]$$

$$= x^3 (5x + 2)^{-4/5} (21x + 8)$$

$$= \dfrac{x^3 (21x + 8)}{(5x + 2)^{4/5}}$$

17. Using the product rule:

$$\dfrac{dy}{dx} = 2x^7 \cdot \dfrac{3}{4}(4x + 1)^{-1/4}(4) + 14x^6 \cdot (4x + 1)^{3/4}$$

$$= 6x^7 (4x + 1)^{-1/4} + 14x^6 (4x + 1)^{3/4}$$

$$= 2x^6 (4x + 1)^{-1/4} [3x + 7(4x + 1)]$$

$$= 2x^6 (4x + 1)^{-1/4} (31x + 7)$$

$$= \dfrac{2x^6 (31x + 7)}{(4x + 1)^{1/4}}$$

19. First rewrite the function as $y = \dfrac{1}{x + 3} = (x + 3)^{-1}$.

Now find the derivative using the general power rule: $\dfrac{dy}{dx} = -1(x + 3)^{-2} = \dfrac{-1}{(x + 3)^2}$

21. First rewrite the function as $y = \dfrac{8}{(x - 4)^3} = 8(x - 4)^{-3}$.

Now find the derivative using the general power rule: $\dfrac{dy}{dx} = -24(x - 4)^{-4} = \dfrac{-24}{(x - 4)^4}$

23. First rewrite the function as $y = \dfrac{3}{(2x + 7)^4} = 3(2x + 7)^{-4}$.

Now find the derivative using the general power rule: $\dfrac{dy}{dx} = -12(2x + 7)^{-5} \cdot 2 = \dfrac{-24}{(2x + 7)^5}$

25. First rewrite the function as $y = \dfrac{\left(x^2+6\right)^7}{x-2} = \left(x^2+6\right)^7 (x-2)^{-1}$. Now find the derivative using the product rule:

$$\frac{dy}{dx} = \left(x^2+6\right)^7\left[-1(x-2)^{-2}\right] + 7\left(x^2+6\right)^6(2x) \cdot (x-2)^{-1}$$

$$= -1\left(x^2+6\right)^7(x-2)^{-2} + 14x\left(x^2+6\right)^6(x-2)^{-1}$$

$$= \left(x^2+6\right)^6(x-2)^{-2}\left[-1\left(x^2+6\right) + 14x(x-2)\right]$$

$$= \left(x^2+6\right)^6(x-2)^{-2}\left(-x^2-6+14x^2-28x\right)$$

$$= \left(x^2+6\right)^6(x-2)^{-2}\left(13x^2-28x-6\right)$$

$$= \frac{\left(x^2+6\right)^6\left(13x^2-28x-6\right)}{(x-2)^2}$$

27. First rewrite the function as $y = \dfrac{3x^2+1}{(x+2)^3} = \left(3x^2+1\right)(x+2)^{-3}$. Now find the derivative using the product rule:

$$\frac{dy}{dx} = \left(3x^2+1\right)\left[-3(x+2)^{-4}\right] + 6x \cdot (x+2)^{-3}$$

$$= -3\left(3x^2+1\right)(x+2)^{-4} + 6x(x+2)^{-3}$$

$$= 3(x+2)^{-4}\left[-1\left(3x^2+1\right) + 2x(x+2)\right]$$

$$= 3(x+2)^{-4}\left(-3x^2-1+2x^2+4x\right)$$

$$= 3(x+2)^{-4}\left(-x^2+4x-1\right)$$

$$= \frac{-3\left(x^2-4x+1\right)}{(x+2)^4}$$

29. First rewrite the function as $y = \left(\dfrac{9x-4}{3x+2}\right)^3 = \dfrac{(9x-4)^3}{(3x+2)^3} = (9x-4)^3(3x+2)^{-3}$.

Now find the derivative using the product rule:

$$\frac{dy}{dx} = (9x-4)^3\left[-3(3x+2)^{-4} \cdot 3\right] + 3(9x-4)^2 \cdot 9(3x+2)^{-3}$$

$$= -9(9x-4)^3(3x+2)^{-4} + 27(9x-4)^2(3x+2)^{-3}$$

$$= 9(9x-4)^2(3x+2)^{-4}\left[-1(9x-4) + 3(3x+2)\right]$$

$$= 9(9x-4)^2(3x+2)^{-4}\left(-9x+4+9x+6\right)$$

$$= 90(9x-4)^2(3x+2)^{-4}$$

$$= \frac{90(9x-4)^2}{(3x+2)^4}$$

31. Finding the expression: $u(x) \cdot v(x) = \left(5x^2+4\right)\left(x^3+11\right) = 5x^5 + 4x^3 + 55x^2 + 44$

33. Finding the expression:
$$h(x) = v(x) \cdot f(x) - u(x) \cdot g(x)$$
$$= (x^2 - 3)(3x^2 + 2) - (x^3 + 5x + 4)(2x)$$
$$= (3x^4 - 9x^2 + 2x^2 - 6) - (2x^4 + 10x^2 + 8x)$$
$$= 3x^4 - 7x^2 - 6 - 2x^4 - 10x^2 - 8x$$
$$= x^4 - 17x^2 - 8x - 6$$

35. Finding the expression:
$$f(u(x)) = f(2x - 9) = (2x - 9)^2 + 3(2x - 9) = 4x^2 - 36x + 81 + 6x - 27 = 4x^2 - 30x + 54$$

37. Yes, the chain rule is appropriate: $\dfrac{dy}{dx} = \dfrac{dy}{du} \cdot \dfrac{du}{dx} = (8)(2) = 16$

39. No, the chain rule is not appropriate: $\dfrac{dy}{da} = 4$

41. The rate of change of y with respect to x is 4 when $x = 2$.

43. Using the chain rule: $\dfrac{dP}{dE} = \dfrac{dP}{dN} \cdot \dfrac{dN}{dE} = (8)(0.6) = 4.8$

The rate of change of population is 4,800 per billion dollars in the economy when the economy is 204 billion dollars.

45. Finding the derivative: $\dfrac{dg}{di} = \dfrac{dg}{du} \cdot \dfrac{du}{di} = \left(\dfrac{6}{5}u^{1/5}\right)\left(-80(i+10)^{-2}\right) = \left(\dfrac{6}{5}\left(\dfrac{80}{i+10}\right)^{1/5}\right)\left(\dfrac{-80}{(i+10)^2}\right) = \dfrac{-96(80)^{1/5}}{(i+10)^{11/5}}$

Evaluating when $i = 7$: $\dfrac{dg}{di}(7) = \dfrac{-96(80)^{1/5}}{(7+10)^{11/5}} \approx -0.453$

The government can expect a decrease in government spending of approximately $453,000,000.

47. **a.** Finding the composition: $(c \circ d)(p) = c(420p^{-2}) = 20 + 10[420p^{-2}]^{-1/3} = 20 + \left(\dfrac{10}{420^{1/3}}\right)p^{2/3}$

Finding the derivative: $\dfrac{dc}{dp} = \dfrac{2}{3}\left(\dfrac{10}{420^{1/3}}\right)p^{-1/3} = \left(\dfrac{20}{3 \cdot 420^{1/3}}\right)p^{-1/3}$

b. Evaluating when $p = 50$: $\dfrac{dc}{dp}(50) = \left(\dfrac{20}{3 \cdot 420^{1/3}}\right)(50)^{-1/3} \approx 0.242$

The cost increases by $0.242 per unit as the price increases from $50 to $51.

49. **a.** Finding the composition: $(N \circ P)(C) = N(1.5C + 0.5) = -(1.5C + 0.5)^2 + 4(1.5C + 0.5) + 12$

Finding the derivative: $\dfrac{dN}{dC} = -2(1.5C + 0.5)(1.5) + 4(1.5) = -4.5C - 1.5 + 6 = -4.5C + 4.5$

b. Evaluating when $C = 1.70$: $\dfrac{dN}{dC}(1.7) = -4.5(1.7) + 4.5 = -3.15$

The manager should expect a decrease of 3,150 pads to be packaged.

51. Finding the derivative: $\dfrac{dD}{dx} = 400 \cdot \dfrac{1}{4}\left(\dfrac{8}{2x-25}\right)^{-3/4} \cdot (-8)(2x-25)^{-2}(2) = \dfrac{-1,600(8)^{-3/4}}{(2x-25)^{5/4}}$

Evaluating when $x = 40$: $\dfrac{dD}{dx}(40) = \dfrac{-1,600(8)^{-3/4}}{(80-25)^{5/4}} \approx -2.246$

The demand will drop by approximately 2,250 tablets.

53. Using the chain rule: $f'(x) = 5(3x^2 + 4)^4(6x) = 30x(3x^2 + 4)^4$

55. Finding the derivative: $f'(x) = \dfrac{2}{3}(x-7)^{-1/3} = \dfrac{2}{3(x-7)^{1/3}}$

Evaluating at $x = 7$ produces a 0 denominator, which is undefined.

57. Solving for y:

$$x^2 + y^2 = 4$$
$$y^2 = 4 - x^2$$
$$y = \pm\sqrt{4 - x^2}$$

59. **a.** Finding the area: $A = \pi(600)^2 = 360,000\pi \text{ ft}^2 \approx 1,130,973 \text{ ft}^2$

b. Evaluating: $2\pi r \dfrac{dr}{dt} = 2\pi(600)(2) = 2,400\pi \approx 7,540$

2.5 Implicit Differentiation

1. Finding the derivative implicitly:

$$\frac{d}{dx}(3y^4) + \frac{d}{dx}(2x^3) = \frac{d}{dx}(4)$$
$$12y^3 y' + 6x^2 = 0$$
$$12y^3 y' = -6x^2$$
$$y' = -\frac{x^2}{2y^3}$$

3. Finding the derivative implicitly:

$$\frac{d}{dx}(3y^3) - \frac{d}{dx}(2y^2) + \frac{d}{dx}(5x^4) - \frac{d}{dx}(x) = \frac{d}{dx}(1)$$
$$9y^2 y' - 4yy' + 20x^3 - 1 = 0$$
$$(9y^2 - 4y)y' = 1 - 20x^3$$
$$y' = \frac{1 - 20x^3}{9y^2 - 4y}$$

5. Finding the derivative implicitly:

$$\frac{d}{dx}(y+6)^8 = \frac{d}{dx}(4x^2) + \frac{d}{dx}(x) - \frac{d}{dx}(4)$$
$$8(y+6)^7 y' = 8x + 1$$
$$y' = \frac{8x+1}{8(y+6)^7}$$

7. Finding the derivative implicitly:

$$\frac{d}{dx}(2y^3 + 5x^2)^3 = \frac{d}{dx}(6x^2) + \frac{d}{dx}(11)$$
$$3(2y^3 + 5x^2)^2 \cdot (6y^2 y' + 10x) = 12x + 0$$
$$18y^2(2y^3 + 5x^2)^2 y' + 30x(2y^3 + 5x^2)^2 = 12x$$
$$18y^2(2y^3 + 5x^2)^2 y' = 12x - 30x(2y^3 + 5x^2)^2$$
$$y' = \frac{2x - 5x(2y^3 + 5x^2)^2}{3y^2(2y^3 + 5x^2)^2}$$

9. Finding the derivative implicitly:

$$\frac{d}{dx}(6x^2 + 4x^2 y^4) = \frac{d}{dx}(7x - 1)$$
$$12x + 4x^2 \cdot 4y^3 y' + 8xy^4 = 7$$
$$16x^2 y^3 y' = 7 - 12x - 8xy^4$$
$$y' = \frac{7 - 12x - 8xy^4}{16x^2 y^3}$$

11. Finding the derivative implicitly:

$$\frac{d}{dx}(5x^2 - 3x^2 y^2 - y^3) = \frac{d}{dx}(2x)$$
$$10x - 3x^2 \cdot 2yy' - 6xy^2 - 3y^2 y' = 2$$
$$(-6x^2 y - 3y^2)y' = 6xy^2 - 10x + 2$$
$$y' = \frac{6xy^2 - 10x + 2}{-6x^2 y - 3y^2}$$

13. Finding the derivative implicitly:

$$\frac{d}{dx}(y+1)^{2/5} = \frac{d}{dx}(1-4x)$$

$$\frac{2}{5}(y+1)^{-3/5}y' = -4$$

$$y' = -10(y+1)^{3/5}$$

15. Finding the derivative implicitly:

$$\frac{d}{dx}(x^2 + y^2) = \frac{d}{dx}(25)$$

$$2x + 2yy' = 0$$

$$y' = -\frac{x}{y}$$

Evaluating at (0,5): $y'(0,5) = -\frac{0}{5} = 0$

17. Finding the derivative implicitly:

$$\frac{d}{dx}(4y^3 + 3x^4 + y) = \frac{d}{dx}(53)$$

$$12y^2y' + 12x^3 + y' = 0$$

$$(12y^2 + 1)y' = -12x^3$$

$$y' = \frac{-12x^3}{12y^2 + 1}$$

Evaluating at (–2,1): $y'(-2,1) = \frac{-12(-2)^3}{12(1)^2 + 1} = \frac{96}{13}$

19. Finding the derivative implicitly:

$$\frac{d}{dx}(x^{-3} + y^{-3}) = \frac{d}{dx}\left(-\frac{35}{216}\right)$$

$$-3x^{-4} - 3y^{-4}y' = 0$$

$$-3y^{-4}y' = 3x^{-4}$$

$$y' = -\frac{y^4}{x^4}$$

Evaluating at (–2,–3): $y'(-2,-3) = -\frac{(-3)^4}{(-2)^4} = -\frac{81}{16}$

21. **a.** Substituting $x = 4$:

$$(4)^2 + y^2 = 25$$

$$16 + y^2 = 25$$

$$y^2 = 9$$

$$y = \pm 3$$

b. Finding the derivative implicitly:

$$\frac{d}{dx}(x^2 + y^2) = \frac{d}{dx}(25)$$

$$2x + 2yy' = 0$$

$$y' = -\frac{x}{y}$$

Evaluating at (4,3): $y'(4,3) = -\frac{4}{3}$

Evaluating at (4,–3): $y'(4,-3) = \frac{4}{3}$

c. Setting the derivative equal to 0:

$$-\frac{x}{y} = 0$$

$$x = 0$$

Substituting to find y:

$$(0)^2 + y^2 = 25$$

$$0 + y^2 = 25$$

$$y^2 = 25$$

$$y = \pm 5$$

The points are $(0,5)$ and $(0,-5)$.

d. Differentiating the area function:

$$\frac{d}{dt}(A) = \frac{d}{dt}(\pi r^2)$$

$$\frac{dA}{dt} = 2\pi r \frac{dr}{dt}$$

$$\frac{dA}{dt} = 2\pi(6)(2) = 24\pi \frac{\text{cm}^2}{\text{s}}$$

23. The error is in step 1:

$$\frac{d}{dx}(4y^3 - 8x^2 + y) = \frac{d}{dx}(0)$$

$$12y^2 y' - 16x + y' = 0$$

25. The error is in step 1:

$$\frac{d}{dx}\left((2y^3 - 5x^2)^4 + 6y\right) = \frac{d}{dx}(0)$$

$$4(2y^3 - 5x^2)^3(6y^2 y' - 10x) + 6y' = 0$$

27. Finding the derivative:

$$\frac{d}{dp}(x^3 + 250p^2) = \frac{d}{dp}(18,000)$$

$$3x^2 \frac{dx}{dp} + 500p = 0$$

$$\frac{dx}{dp} = \frac{-500p}{3x^2}$$

Evaluating at $(4, 24.101)$: $\dfrac{dx}{dp}(4, 24.101) = \dfrac{-500(4)}{3(24.101)^2} \approx -1.15$

The approximate decrease is 1,150 collars.

29. Finding the derivative: $\dfrac{dC}{dt} = \dfrac{dC}{dx} \cdot \dfrac{dx}{dt} = \left(20x^{1/3} + 10x^{-1/3}\right)\dfrac{dx}{dt}$

Evaluating when $x = 1,728$ and $\dfrac{dx}{dt} = 350$: $\dfrac{dC}{dt} = \left(20 \cdot 1,728^{1/3} + 10 \cdot 1,728^{-1/3}\right)(350) \approx 84,292$

The cost is increasing \$84,292 per month.

31. Finding the derivative: $\dfrac{dR}{dt} = \dfrac{dR}{dx} \cdot \dfrac{dx}{dt} = \dfrac{25}{2}(3.5x^2 + 25x)^{-1/2} \cdot (7x + 25)\dfrac{dx}{dt} = 12.5(3.5x^2 + 25x)^{-1/2}(7x + 25)\dfrac{dx}{dt}$

Evaluating when $x = 750$ and $\dfrac{dx}{dt} = 10$: $\dfrac{dR}{dt} = 12.5(3.5 \cdot 750^2 + 25 \cdot 750)^{-1/2}(7 \cdot 750 + 25)(10) \approx 467.71$

The monthly revenue is increasing by approximately \$467.71.

33. Finding the derivative implicitly:

$$\frac{d}{dx}\left(2y^3 + 3x^2\right) = \frac{d}{dx}(25)$$

$$6y^2 y' + 6x = 0$$

$$y' = -\frac{6x}{6y^2} = -\frac{x}{y^2}$$

35. Finding the derivative implicitly:

$$\frac{d}{dx}\left(x^3 + 3y^3 + y\right) = \frac{d}{dx}(92)$$

$$3x^2 + 9y^2 y' + y' = 0$$

$$\left(9y^2 + 1\right)y' = -3x^2$$

$$y' = \frac{-3x^2}{9y^2 + 1}$$

Evaluating at (2,3): $y'(2,3) = \dfrac{-3(2)^2}{9(3)^2 + 1} = -\dfrac{6}{41}$

At the point (2,3), the slope of the tangent line is $-\dfrac{6}{41}$.

37. Finding the implicit derivative:

$$\frac{d}{dt}\left(PV^{1.4}\right) = \frac{d}{dt}(C)$$

$$1.4 PV^{0.4}\frac{dV}{dt} + V^{1.4}\frac{dP}{dt} = 0$$

$$1.4(30)(500)^{0.4}(-10) + 500^{1.4}\frac{dP}{dt} = 0$$

$$\frac{dP}{dt} = \frac{420(500)^{0.4}}{500^{1.4}} = 0.84$$

The pressure is increasing at the rate of 0.84 lb/in^2 per second.

39. Finding the derivative:

$$\frac{dC}{dt} = \frac{dC}{dp} \cdot \frac{dp}{dt}$$

$$= 13{,}500\left(\frac{250p + 20}{80p + 90}\right)^{-1/4} \cdot \left(\frac{(80p + 90)\cdot 250 - (250p + 20)\cdot 80}{(80p + 90)^2}\right)\frac{dp}{dt}$$

$$= 13{,}500\left(\frac{250p + 20}{80p + 90}\right)^{-1/4}\left(\frac{20{,}900}{(80p + 90)^2}\right)\frac{dp}{dt}$$

Evaluating when $p = 0.40$ and $\dfrac{dp}{dt} = 0.02$: $\dfrac{dC}{dt} = 13{,}500\left(\dfrac{250(0.40) + 20}{80(0.40) + 90}\right)^{-1/4}\left(\dfrac{20{,}900}{(80(0.40) + 90)^2}\right)(0.02) \approx 381$

The cost is increasing by approximately $381 per month.

41. Substituting $x = 0$:

$$f(x) = \frac{5(0)+2}{3(0)-4} = -\frac{1}{2}$$

Substituting $f(x) = 0$:

$$\frac{5x+2}{3x-4} = 0$$
$$5x+2 = 0$$
$$5x = -2$$
$$x = -\frac{2}{5}$$

The intercepts are $\left(0,-\frac{1}{2}\right)$ and $\left(-\frac{2}{5},0\right)$.

43. Finding the limit: $\displaystyle\lim_{x\to\infty}\frac{5x+2}{3x-4} = \lim_{x\to\infty}\frac{5+\dfrac{2}{x}}{3-\dfrac{4}{x}} = \frac{5}{3}$

45. Solving the inequality:
$$6x - 6 > 0$$
$$6x > 6$$
$$x > 1$$
The solution is the interval $(1,\infty)$.

47. Finding the first and second derivatives:
$$f'(x) = \frac{(3x-4)\cdot 5 - (5x+2)\cdot 3}{(3x-4)^2} = \frac{15x-20-15x-6}{(3x-4)^2} = \frac{-26}{(3x-4)^2}$$
$$f''(x) = -26\left(-2(3x-4)^{-3}\right)\cdot 3 = \frac{156}{(3x-4)^3}$$

Chapter 2 Test

1. Finding the derivative: $f'(x) = 3x^2 + 5$

2. Finding the derivative: $f'(x) = 0$

3. Finding the derivative: $f'(x) = \frac{4}{3}(5-3x)^{1/3}(-3) = -4(5-3x)^{1/3}$

4. Finding the derivative: $f'(x) = 3 - 8x + 3x^2 = 3x^2 - 8x + 3$

5. Finding the derivative: $f'(x) = (2x+1)\cdot 3 + 2(3x-8) = 6x+3+6x-16 = 12x-13$

6. Finding the derivative:
$$f'(x) = (3x^2)\cdot 3(5x+4)^2 \cdot 5 + 6x(5x+4)^3 = 3x(5x+4)^2(15x+2(5x+4)) = 3x(5x+4)^2(25x+8)$$

7. Finding the derivative:
$$f'(x) = (x-4)^2 \cdot 3(x+1)^2 + 2(x-4)(x+1)^3$$
$$= (x-4)(x+1)^2(3(x-4)+2(x+1))$$
$$= (x-4)(x+1)^2(5x-10)$$
$$= 5(x-4)(x+1)^2(x-2)$$

8. First write the function as $f(x) = 2(3x+4)^{-1/3}$. Finding the derivative:
$$f'(x) = -\frac{2}{3}(3x+4)^{-4/3}(3) = -2(3x+4)^{-4/3} = \frac{-2}{(3x+4)^{4/3}}$$

9. Finding the derivative: $f'(x) = \dfrac{(x+3)(1)-(x+1)(1)}{(x+3)^2} = \dfrac{x+3-x-1}{(x+3)^2} = \dfrac{2}{(x+3)^2}$

10. Finding the derivative:

$$f'(x) = \dfrac{(x-4)^2 \cdot 3(x+2)^2 - (x+2)^3 \cdot 2(x-4)}{(x-4)^4}$$

$$= \dfrac{(x-4)(x+2)^2 \left[3(x-4)-2(x+2)\right]}{(x-4)^4}$$

$$= \dfrac{(x+2)^2 (3x-12-2x-4)}{(x-4)^3}$$

$$= \dfrac{(x+2)^2 (x-16)}{(x-4)^3}$$

11. Any trinomial with degree 3 (or less) will have a derivative equal to 0.

12. Using the chain rule: $\dfrac{dy}{dx} = \dfrac{dy}{du} \cdot \dfrac{du}{dx} = 3 \cdot 1 = 3$

13. Using the chain rule: $\dfrac{dy}{dx} = \dfrac{dy}{du} \cdot \dfrac{du}{dx} = 2u \cdot 2 = 4(2x+3) = 8x+12$

14. Using implicit differentiation:

$$\dfrac{d}{dx}\left(x^2 + 3y^3 + 4\right) = \dfrac{d}{dx}(0)$$

$$2x + 9y^2 y' = 0$$

$$y' = \dfrac{-2x}{9y^2}$$

15. Using implicit differentiation:

$$\dfrac{d}{dx}\left(5x - y^3 + x^2 y^2\right) = \dfrac{d}{dx}(10)$$

$$5 - 3y^2 y' + 2x^2 yy' + 2xy^2 = 0$$

$$\left(2x^2 y - 3y^2\right)y' = -2xy^2 - 5$$

$$y' = \dfrac{-2xy^2 - 5}{2x^2 y - 3y^2}$$

16. a. Finding the first derivative: $f'(x) = 12x^2 + 6x - 1$

    b. Finding the second derivative: $f''(x) = 24x + 6$

    c. Finding the third derivative: $f'''(x) = 24$

17. a. Finding the first derivative: $f'(x) = 3(x-3)^2$

    b. Finding the second derivative: $f''(x) = 6(x-3)$

18. a. Evaluating the function: $f(2) = 3(2)^2 - 2 = 10$

    b. Finding the derivative: $f'(x) = 6x - 1$

       Evaluating the derivative: $f'(2) = 6(2) - 1 = 11$

19. Finding the derivative: $f'(x) = x^3(2x) + 3x^2(x^2+1) = 2x^4 + 3x^4 + 3x^2 = 5x^4 + 3x^2$

    Evaluating when $x = 0$: $f'(0) = 5(0)^4 + 3(0)^2 = 0$

20. Finding the derivative: $f'(x) = 14x + 21$

    Setting the derivative equal to 0:

    $$14x + 21 = 0$$
    $$14x = -21$$
    $$x = -\dfrac{3}{2}$$

21. Finding the derivative: $\dfrac{dy}{dx} = 2(x-3)$

Evaluating when $x = 4$: $\dfrac{dy}{dx}(4) = 2(4-3) = 2$

Using the point-slope formula with the point (4,3):
$$y - 3 = 2(x-4)$$
$$y - 3 = 2x - 8$$
$$y = 2x - 5$$

22. An 18 mm thick cup has a breaking strength of 10 lb/in^2. An increase of 1 mm in thickness will result in an increase of 0.8 lb/in^2 in breaking strength.

23. When the grain size is 8 $(d/um)^{1/2}$ the transition temperature is 50° and is increasing, but at a decreasing rate.

24. **a.** Finding the derivative: $f'(x) = -\dfrac{7}{320}x + \dfrac{7}{4}$

   b. Evaluating when $x = 80$: $f'(80) = -\dfrac{7}{320}(80) + \dfrac{7}{4} = -\dfrac{7}{4} + \dfrac{7}{4} = 0$

   The human cannonball is at his maximum height when he is 80 horizontal feet from the front of the cannon.

25. Finding the rate equation and evaluating when $r = 4$ and $\dfrac{dr}{dt} = 2$:

$$\frac{d}{dt}(A) = \frac{d}{dt}(\pi r^2)$$
$$\frac{dA}{dt} = 2\pi r \frac{dr}{dt}$$
$$\frac{dA}{dt} = 2\pi(4)(2) = 16\pi \approx 50.3$$

The area is increasing by 50.3 cm^2/min.

26. Finding the derivative: $C'(x) = -203{,}000x^{-2} + 2 = 2 - \dfrac{203{,}000}{x^2}$

Evaluating when $x = 317$: $C'(317) = 2 - \dfrac{203{,}000}{317^2} \approx -0.02$

The manager can expect the inventory costs to decrease by $0.02.

27. Finding the derivative: $P'(t) = \dfrac{(t^2 + 20)(1700t) - (850t^2)(2t)}{(t^2 + 20)^2} = \dfrac{1700t^3 + 34000t - 1700t^3}{(t^2 + 20)^2} = \dfrac{34000t}{(t^2 + 20)^2}$

Evaluating when $t = 10$: $P'(10) = \dfrac{34000(10)}{(10^2 + 20)^2} \approx 23.61$

Evaluating when $t = 11$: $P'(11) = \dfrac{34000(11)}{(11^2 + 20)^2} \approx 18.81$

At year 10 the marginal profit is approximately $23,610 per year, and at year 11 the marginal profit is approximately $18,810 per year.

Chapter 3
Applying the Derivative

3.1 The First Derivative and the Behavior of Functions

1. Constructing a sign chart:

(+)	(−)	(+)	(−)	(−)	(+)	

$\overset{\text{(+)}}{\underset{1}{\bullet}}\ \overset{\text{(−)}}{\underset{3}{\bullet}}\ \overset{\text{(+)}}{\underset{5}{\bullet}}\ \overset{\text{(−)}}{\underset{6}{\bullet}}\ \overset{\text{(−)}}{\underset{7}{\bullet}}\ \overset{\text{(+)}}{}\ \longrightarrow x$

3. Constructing a sign chart:

$\overset{\text{(−)}}{}\ \overset{\text{(+)}}{\underset{4}{\bullet}}\ \overset{\text{(+)}}{\underset{6}{\bullet}}\ \overset{\text{(−)}}{\underset{7}{\bullet}}\ \overset{\text{(−)}}{\underset{9}{\bullet}}\ \longrightarrow x$

5. Constructing a summary table:

Interval/Value	$f'(x)$	$f(x)$	Behavior of $f(x)$
$(-\infty,5)$	$(-)$		$f(x)$ is decreasing
5	0	4	$(5,4)$ is a relative and absolute minimum
$(5,\infty)$	$(+)$		$f(x)$ is increasing

$f(x)$ has no absolute maximum

7. Constructing a summary table:

Interval/Value	$f'(x)$	$f(x)$	Behavior of $f(x)$
0		16	$(0,16)$ is an absolute maximum
$(0,11)$	$(-)$		$f(x)$ is decreasing
11	0	3	$(11,3)$ is a relative and absolute minimum
$(11,\infty)$	$(+)$		$f(x)$ is increasing

$y = 10$ is a horizontal asymptote

9. Finding the derivative: $f'(x) = 3$

Since the derivative cannot equal 0 and is not undefined, there are no critical values for the function.

11. Finding the derivative: $f'(x) = 4x - 3$

Setting the derivative equal to 0:

$$4x - 3 = 0$$
$$4x = 3$$
$$x = \frac{3}{4}$$

The only critical value for the function is $x = \frac{3}{4}$.

13. Finding the derivative: $f'(x) = x^3 + 3x^2 - 4x$

Setting the derivative equal to 0:

$$x^3 + 3x^2 - 4x = 0$$
$$x(x^2 + 3x - 4) = 0$$
$$x(x + 4)(x - 1) = 0$$
$$x = -4, 0, 1$$

The critical values for the function are $x = -4, 0, 1$.

15. Finding the derivative: $f'(x) = 4x^3 + 12x^2$

Setting the derivative equal to 0:

$$4x^3 + 12x^2 = 0$$
$$4x^2(x + 3) = 0$$
$$x = -3, 0$$

The critical values for the function are $x = -3, 0$.

17. Finding the derivative: $f'(x) = \dfrac{(x-5)(0) - 1(1)}{(x-5)^2} = -\dfrac{1}{(x-5)^2}$

Since the derivative cannot equal 0 and is not undefined ($x = 5$ is not in the domain of the function), there are no critical values for the function.

19. Finding the derivative: $f'(x) = \dfrac{(x-3)(4) - 4x(1)}{(x-3)^2} = \dfrac{4x - 12 - 4x}{(x-3)^2} = -\dfrac{12}{(x-3)^2}$

Since the derivative cannot equal 0 and is not undefined ($x = 3$ is not in the domain of the function), there are no critical values for the function.

21. Finding the derivative: $f'(x) = \dfrac{(x+2)(2x) - x^2(1)}{(x+2)^2} = \dfrac{2x^2 + 4x - x^2}{(x+2)^2} = \dfrac{x^2 + 4x}{(x+2)^2} = \dfrac{x(x+4)}{(x+2)^2}$

The derivative equals 0 when $x = -4, 0$, and is not undefined ($x = -2$ is not in the domain of the function). The critical values are $x = -4, 0$.

23. Finding the derivative: $f'(x) = \dfrac{3}{5} x^{-2/5} = \dfrac{3}{5x^{2/5}}$

The derivative does not equal 0, and is undefined when $x = 0$. The critical value is $x = 0$.

25. Finding the derivative: $f'(x) = 8$

Since the derivative is always positive, the function is increasing on $(-\infty, \infty)$.

27. Finding the derivative: $f'(x) = 9x^2 - 24x = 3x(3x - 8)$

Constructing a summary table:

Interval/Value	$f'(x)$	$f(x)$	Behavior of $f(x)$
$(-\infty, 0)$	$(+)$		$f(x)$ is increasing
0	0	4	$(0, 4)$ is a relative maximum
$\left(0, \dfrac{8}{3}\right)$	$(-)$		$f(x)$ is decreasing
$\dfrac{8}{3}$	0	$-\dfrac{256}{9}$	$\left(\dfrac{8}{3}, -\dfrac{256}{9}\right)$ is a relative minimum
$\left(\dfrac{8}{3}, \infty\right)$	$(+)$		$f(x)$ is increasing

The function is increasing on the intervals $(-\infty, 0)$ and $\left(\dfrac{8}{3}, \infty\right)$, and decreasing on the interval $\left(0, \dfrac{8}{3}\right)$.

29. Finding the derivative: $f'(x) = 8x^3 + 24x^2 = 8x^2(x+3)$

Constructing a summary table:

Interval/Value	$f'(x)$	$f(x)$	Behavior of $f(x)$
$(-\infty,-3)$	$(-)$		$f(x)$ is decreasing
-3	0	-54	$(-3,-54)$ is a relative minimum
$(-3,0)$	$(+)$		$f(x)$ is increasing
0	0	0	$(0,0)$ is a horizontal tangent
$(0,\infty)$	$(+)$		$f(x)$ is increasing

The function is increasing on the intervals $(-3,0)$ and $(0,\infty)$, and decreasing on the interval $(-\infty,-3)$.

31. Finding the derivative: $f'(x) = \dfrac{1}{3}x^{-2/3} = \dfrac{1}{3x^{2/3}}$

Since the derivative is always positive, the function is increasing on $(-\infty,0)$ and $(0,\infty)$. Note that $x = 0$ is excluded, since the derivative is undefined when $x = 0$.

33. Finding the derivative: $f'(x) = \dfrac{2}{5}(x-6)^{-3/5} = \dfrac{2}{5(x-6)^{3/5}}$

Constructing a summary table:

Interval/Value	$f'(x)$	$f(x)$	Behavior of $f(x)$
$(-\infty,6)$	$(-)$		$f(x)$ is decreasing
6	undefined	-3	$(6,-3)$ is a relative minimum
$(6,\infty)$	$(+)$		$f(x)$ is increasing

The function is decreasing on the interval $(-\infty,6)$, and increasing on the interval $(6,\infty)$.

35. Write the function as $f(x) = (2x-10)^{1/2}$. Finding the derivative: $f'(x) = \dfrac{1}{2}(2x-10)^{-1/2} \cdot 2 = \dfrac{1}{(2x-10)^{1/2}}$

Constructing a summary table (note that the domain is $x \geq 5$):

Interval/Value	$f'(x)$	$f(x)$	Behavior of $f(x)$
5	undefined	0	$(5,0)$ is a relative minimum
$(5,\infty)$	$(+)$		$f(x)$ is increasing

The function is increasing on the interval $(5,\infty)$.

37. Finding the derivative: $f'(x) = 2x + 12 = 2(x+6)$

Constructing a summary table:

Interval/Value	$f'(x)$	$f(x)$	Behavior of $f(x)$
$(-\infty,-6)$	$(-)$		$f(x)$ is decreasing
-6	0	-36	$(-6,-36)$ is a relative minimum
$(-6,\infty)$	$(+)$		$f(x)$ is increasing

The function has a relative minimum at $(-6,-36)$.

39. Finding the derivative: $f'(x) = 32 - 8x = 8(4-x)$

Constructing a summary table:

Interval/Value	$f'(x)$	$f(x)$	Behavior of $f(x)$
$(-\infty, 4)$	$(+)$		$f(x)$ is decreasing
4	0	64	$(4, 64)$ is a relative maximum
$(4, \infty)$	$(-)$		$f(x)$ is increasing

The function has a relative maximum at $(4, 64)$.

41. Finding the derivative: $f'(x) = 5x^4 - 30x^3 = 5x^3(x-6)$

Constructing a summary table:

Interval/Value	$f'(x)$	$f(x)$	Behavior of $f(x)$
$(-\infty, 0)$	$(+)$		$f(x)$ is increasing
0	0	10	$(0, 10)$ is a relative maximum
$(0, 6)$	$(-)$		$f(x)$ is decreasing
6	0	-1934	$(6, -1934)$ is a relative minimum
$(6, \infty)$	$(+)$		$f(x)$ is increasing

The function has a relative maximum at $(0, 10)$ and a relative minimum at $(6, -1934)$.

43. Write the function as $f(x) = x + 4x^{-1}$. Finding the derivative: $f'(x) = 1 - 4x^{-2} = 1 - \dfrac{4}{x^2} = \dfrac{x^2 - 4}{x^2} = \dfrac{(x+2)(x-2)}{x^2}$

Constructing a summary table:

Interval/Value	$f'(x)$	$f(x)$	Behavior of $f(x)$
$(-\infty, -2)$	$(+)$		$f(x)$ is increasing
-2	0	-4	$(-2, -4)$ is a relative maximum
$(-2, 0)$	$(-)$		$f(x)$ is decreasing
0	undefined	undefined	$x = 0$ is a vertical asymptote
$(0, 2)$	$(-)$		$f(x)$ is decreasing
2	0	4	$(2, 4)$ is a relative minimum
$(2, \infty)$	$(+)$		$f(x)$ is increasing

The function has a relative maximum at $(-2, -4)$ and a relative minimum at $(2, 4)$.

45. Write the function as $f(x) = (x-7)^{-1}$. Finding the derivative: $f'(x) = -(x-7)^{-2} = -\dfrac{1}{(x-7)^2}$

Constructing a summary table:

Interval/Value	$f'(x)$	$f(x)$	Behavior of $f(x)$
$(-\infty, 7)$	$(-)$		$f(x)$ is decreasing
7	undefined	undefined	$x = 7$ is a vertical asymptote
$(7, \infty)$	$(-)$		$f(x)$ is decreasing

The function does not have any relative extrema.

47. Finding the derivative: $f'(x) = 3x^2 - 8x - 3 = (3x+1)(x-3)$

The critical values are $x = -\dfrac{1}{3}, 3$, but only $x = 3$ lies in the required interval $[0,5]$. Evaluating the function at the critical values and the endpoints of the interval:

x	$f(x)$	Behavior of $f(x)$
0	1	neither maximum nor minimum
3	−17	absolute minimum
5	11	absolute maximum

The absolute maximum occurs at $(5,11)$ and the absolute minimum occurs at $(3,-17)$.

49. Write the function as $f(x) = 2(x-1)^{-1}$. Finding the derivative: $f'(x) = -2(x-1)^{-2} = -\dfrac{2}{(x-1)^2}$

This function is decreasing everywhere, so the absolute maximum will occur when $x = 3$ and there will be no absolute minimum on the interval $[3, \infty)$. The absolute maximum occurs at $(3,1)$, and there is no absolute minimum.

51. Based on the graph, the function will be minimized when $r = 2$ inches. The minimum cost is approximately \$300.

53. Based on the graph, a width of 150 feet will maximize the area. The maximum area is approximately 45,000 square feet.

55. Based on the graph, the area of the page is maximized when $w = 7$ inches. The maximum area is approximately 51 square inches.

57. Based on the graph, the percentage P will be maximized when $t = 3.9$ weeks. The maximum value is approximately 6.2%

59. Finding the derivative: $V'(t) = \dfrac{(t+3)(340) - 340t(1)}{(t+3)^2} = \dfrac{340t + 1020 - 340t}{(t+3)^2} = \dfrac{1020}{(t+3)^2}$

This function is increasing everywhere.

a. The claim is false. Since the function is increasing everywhere, the speed will continue to increase.

b. Since the speed continues to increase, its maximum will occur after $t = 4$ seconds. The maximum speed is

$$V(4) = \dfrac{340(4)}{4+3} = \dfrac{1360}{7} \approx 194.29 \text{ mph.}$$

61. Finding the derivative: $C'(x) = \dfrac{(180x)(2x) - (144 + x^2)(180)}{(180x)^2} = \dfrac{180(2x^2 - 144 - x^2)}{(180x)^2} = \dfrac{x^2 - 144}{180x^2}$

This function is decreasing for $x < 12$ and increasing for $x > 12$.

a. The cost will be minimized when $x = 12$. The manufacturer must produce 1,200 PoE systems to minimize its cost.

b. Since the cost is increasing for $x > 12$, the cost will continue to increase.

63. Write the function as $S(x) = \dfrac{1}{2}x(300-x)^{1/2}$. Finding the derivative:

$$S'(x) = \dfrac{1}{2}x \cdot \dfrac{1}{2}(300-x)^{-1/2}(-1) + \dfrac{1}{2}(300-x)^{1/2}$$

$$= \dfrac{1}{2}(300-x)^{-1/2}\left(-\dfrac{1}{2}x + 300 - x\right)$$

$$= \dfrac{1}{2}(300-x)^{-1/2}\left(300 - \dfrac{3}{2}x\right)$$

$$= \dfrac{3}{4}(300-x)^{-1/2}(200-x)$$

This function is increasing for $x < 200$ and decreasing for $x > 200$.

a. The strength will be maximized when $x = 200$. The strength will be maximum when 200 milligrams have been administered.

b. Since the strength is decreasing for $x > 200$, once the reaction to the drug is maximized, it will only decrease after that dosage.

65. Write the function as $R(p) = (d+c)p^{-1}$. Finding the derivative: $R'(p) = -(d+c)p^{-2} = -\dfrac{d+c}{p^2}$

Since d and c are positive, this derivative is always negative and thus the expected return is always decreasing.

67. Write the function as $M(d) = \dfrac{20}{4}d^{-2} = 5d^{-2}$. Finding the derivative: $M'(d) = -10d^{-3} = -\dfrac{10}{d^3}$

This derivative is always negative for $d > 0$, and thus the modulus of rupture decreases as d increases.

69. Finding the derivative: $f'(x) = 3x^2 - 4$

Setting the derivative equal to 0:

$$3x^2 - 4 = 0$$
$$3x^2 = 4$$
$$x^2 = \frac{4}{3}$$
$$x = \pm\frac{2}{\sqrt{3}}$$

Constructing a summary table:

Interval/Value	$f'(x)$	$f(x)$	Behavior of $f(x)$
$\left(-\infty, -\dfrac{2}{\sqrt{3}}\right)$	(+)		$f(x)$ is increasing
$-\dfrac{2}{\sqrt{3}}$	0	$5 + \dfrac{16}{3\sqrt{3}}$	$\left(-\dfrac{2}{\sqrt{3}}, 5 + \dfrac{16}{3\sqrt{3}}\right)$ is a relative maximum
$\left(-\dfrac{2}{\sqrt{3}}, \dfrac{2}{\sqrt{3}}\right)$	(−)		$f(x)$ is decreasing
$\dfrac{2}{\sqrt{3}}$	0	$5 - \dfrac{16}{3\sqrt{3}}$	$\left(\dfrac{2}{\sqrt{3}}, 5 - \dfrac{16}{3\sqrt{3}}\right)$ is a relative minimum
$\left(\dfrac{2}{\sqrt{3}}, \infty\right)$	(+)		$f(x)$ is increasing

The function has a relative maximum at $\left(-\dfrac{2}{\sqrt{3}}, 5 + \dfrac{16}{3\sqrt{3}}\right)$ $(0,0)$ and a relative minimum at $\left(\dfrac{2}{\sqrt{3}}, 5 - \dfrac{16}{3\sqrt{3}}\right)$.

71. Finding the derivative:

$$C'(t) = \frac{(t^2 + 6t + 9)(0.02) - (0.02t)(2t + 6)}{(t^2 + 6t + 9)^2}$$

$$= \frac{0.02(t^2 + 6t + 9 - 2t^2 - 6t)}{(t + 3)^4}$$

$$= \frac{0.02(9 - t^2)}{(t + 3)^4}$$

$$= \frac{0.02(3 + t)(3 - t)}{(t + 3)^4}$$

$$= \frac{0.02(3 - t)}{(t + 3)^3}$$

This function is increasing for $t < 3$ and decreasing for $t > 3$. The concentration will reach a maximum when $t = 3$ hours. The maximum concentration is: $C(3) = \dfrac{0.02(3)}{(3)^2 + 6(3) + 9} = \dfrac{0.06}{36} = \dfrac{6}{3600} = \dfrac{1}{600}$ mg/cm^3

73. Finding the derivative: $f'(x) = -3x^2 + 18x$

75. First write the function as $f(x) = \dfrac{3}{5}x^{-2/5}$. Finding the derivative: $f'(x) = -\dfrac{6}{25}x^{-7/5} = -\dfrac{6}{25x^{7/5}}$

77. Finding the derivative: $f'(x) = \dfrac{(3x-4)(5)-(5x+2)(3)}{(3x-4)^2} = \dfrac{15x-20-15x-6}{(3x-4)^2} = -\dfrac{26}{(3x-4)^2}$

79. The function is undefined when $x = \dfrac{4}{3}$.

81. Finding the first and second derivatives:
$$f'(x) = 3x^2 + 9x - 12$$
$$f''(x) = 6x + 9$$

3.2 The Second Derivative and the Behavior of Functions

1. Sketching the graph:

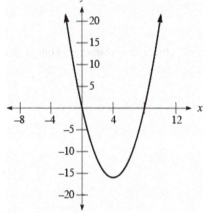

3. Sketching the graph:

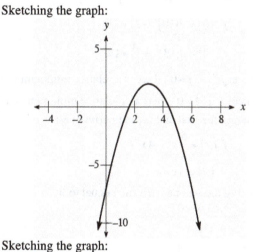

5. Sketching the graph:

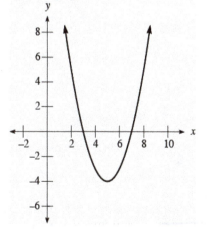

7. Sketching the graph:

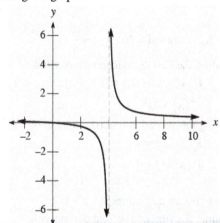

9. Sketching the graph:

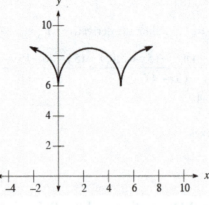

11. Applying the second derivative test:

$$f''(-3) = 3\left((-3)^2 - 3\right) = 18$$

$$f''(0) = 3\left((0)^2 - 3\right) = -9$$

$$f''(3) = 3\left((3)^2 - 3\right) = 18$$

Since $f''(-3) > 0$, there is a relative minimum at $x = -3$. Since $f''(0) < 0$, there is a relative maximum at $x = 0$.
Since $f''(3) > 0$, there is a relative minimum at $x = 3$.

13. Finding the first and second derivatives:

$$f'(x) = \frac{5}{2}x^2 - 4x$$

$$f''(x) = 5x - 4$$

Setting the second derivative equal to 0:

$$5x - 4 = 0$$
$$5x = 4$$
$$x = \frac{4}{5}$$

The hypercritical value is $x = \frac{4}{5}$.

15. Finding the first and second derivatives:

$$f'(x) = x^3 - 6x^2 + 1$$

$$f''(x) = 3x^2 - 12x$$

Setting the second derivative equal to 0:

$$3x^2 - 12x = 0$$
$$3x(x - 4) = 0$$
$$x = 0, 4$$

The hypercritical values are $x = 0, 4$.

17. Finding the first and second derivatives:

$$f'(x) = \frac{2}{3}x^{-1/3}$$

$$f''(x) = -\frac{2}{9}x^{-4/3} = -\frac{2}{9x^{4/3}}$$

The second derivative cannot equal 0, but it is undefined when $x = 0$. The hypercritical value is $x = 0$.

19. Finding the first and second derivatives:

$$f'(x) = \frac{(x+1)(1)-(x-1)(1)}{(x+1)^2} = \frac{x+1-x+1}{(x+1)^2} = \frac{2}{(x+1)^2} = 2(x+1)^{-2}$$

$$f''(x) = -4(x+1)^{-3} = -\frac{4}{(x+1)^3}$$

The second derivative cannot equal 0, but it is undefined when $x = -1$. Since this value is not in the domain of the function, there are no hypercritical values.

21. Constructing a summary table for the second derivative:

Interval/Value	$f''(x)$	Behavior of $f(x)$
$(-\infty,0)$	$(-)$	$f(x)$ is concave downward
0	0	$x = 0$ is an inflection point
$(0,\infty)$	$(+)$	$f(x)$ is concave upward

The function is concave downward on $(-\infty,0)$, and concave upward on $(0,\infty)$.

23. Setting the second derivative equal to 0:

$$6(x^2-1) = 0$$
$$6(x+1)(x-1) = 0$$
$$x = -1,1$$

Constructing a summary table for the second derivative:

Interval/Value	$f''(x)$	Behavior of $f(x)$
$(-\infty,-1)$	$(+)$	$f(x)$ is concave upward
-1	0	$x = -1$ is an inflection point
$(-1,1)$	$(-)$	$f(x)$ is concave downward
1	0	$x = 1$ is an inflection point
$(1,\infty)$	$(+)$	$f(x)$ is concave upward

The function is concave downward on $(-1,1)$, and concave upward on $(-\infty,-1)$ and $(1,\infty)$.

25. Constructing a summary table for the second derivative:

Interval/Value	$f''(x)$	Behavior of $f(x)$
$(-\infty,-1)$	$(+)$	$f(x)$ is concave upward
-1	undefined	$x = -1$ is a vertical asymptote
$(-1,\infty)$	$(-)$	$f(x)$ is concave downward

The function is concave downward on $(-1,\infty)$, and concave upward on $(-\infty,-1)$.

27. Finding the first and second derivatives:

$$f'(x) = 2x-4$$
$$f''(x) = 2$$

The second derivative is always positive, so the function is concave upward on $(-\infty,\infty)$, and never concave downward.

29. Finding the first and second derivatives:
$$f'(x) = 3x^2 - 6x$$
$$f''(x) = 6x - 6$$
Setting the second derivative equal to 0:
$$6x - 6 = 0$$
$$6x = 6$$
$$x = 1$$
Constructing a summary table for the second derivative:

Interval/Value	$f''(x)$	Behavior of $f(x)$
$(-\infty, 1)$	$(-)$	$f(x)$ is concave downward
1	0	$x = 1$ is an inflection point
$(1, \infty)$	$(+)$	$f(x)$ is concave upward

The function is concave downward on $(-\infty, 1)$, and concave upward on $(1, \infty)$.

31. Finding the first and second derivatives:
$$f'(x) = 3(1-x)^2(-1) = -3(1-x)^2$$
$$f''(x) = -6(1-x)(-1) = 6(1-x)$$
Setting the second derivative equal to 0:
$$6(1-x) = 0$$
$$x = 1$$
Constructing a summary table for the second derivative:

Interval/Value	$f''(x)$	Behavior of $f(x)$
$(-\infty, 1)$	$(+)$	$f(x)$ is concave upward
1	0	$x = 1$ is an inflection point
$(1, \infty)$	$(-)$	$f(x)$ is concave downward

The function is concave upward on $(-\infty, 1)$, and concave downward on $(1, \infty)$.

33. Finding the first and second derivatives:
$$f'(x) = \frac{1}{3}(x+1)^{-2/3}$$
$$f''(x) = -\frac{2}{9}(x+1)^{-5/3}$$
The second derivative is never equal to 0, but it is undefined when $x = -1$.
Constructing a summary table for the second derivative:

Interval/Value	$f''(x)$	Behavior of $f(x)$
$(-\infty, -1)$	$(+)$	$f(x)$ is concave upward
-1	undefined	$x = -1$ is an inflection point
$(-1, \infty)$	$(-)$	$f(x)$ is concave downward

The function is concave upward on $(-\infty, -1)$, and concave downward on $(-1, \infty)$.

35. First write the function as $f(x) = 10(x^2 + 3)^{-1}$. Finding the first and second derivatives:

$$f'(x) = -10(x^2 + 3)^{-2}(2x) = -\frac{20x}{(x^2 + 3)^2}$$

$$f''(x) = -\frac{(x^2 + 3)^2(20) - (20x) \cdot 2(x^2 + 3)(2x)}{(x^2 + 3)^4}$$

$$= -\frac{20(x^2 + 3)(x^2 + 3 - 4x^2)}{(x^2 + 3)^4}$$

$$= -\frac{20(3 - 3x^2)}{(x^2 + 3)^3}$$

$$= -\frac{60(1 + x)(1 - x)}{(x^2 + 3)^3}$$

Constructing a summary table for the second derivative:

Interval/Value	$f''(x)$	Behavior of $f(x)$
$(-\infty, -1)$	$(+)$	$f(x)$ is concave upward
-1	0	$x = -1$ is an inflection point
$(-1, 1)$	$(-)$	$f(x)$ is concave downward
1	0	$x = 1$ is an inflection point
$(1, \infty)$	$(+)$	$f(x)$ is concave upward

The function is concave upward on $(-\infty, -1)$ and $(1, \infty)$, and concave downward on $(-1, 1)$.

37. Finding the first and second derivatives:

$$R'(x) = 20x - 2x^2$$
$$R''(x) = 20 - 4x = 4(5 - x)$$

Constructing a summary table for the second derivative:

Interval/Value	$R''(x)$	Behavior of $R(x)$
$(0, 5)$	$(+)$	$R(x)$ is concave upward
5	0	$x = 5$ is an inflection point
$(5, 10)$	$(-)$	$R(x)$ is concave downward

The function is concave upward on $(0, 5)$, and concave downward on $(5, 10)$. The point of diminishing returns is $\$5,00,000$ spent on advertising.

39. Finding the first and second derivatives:

$$R'(x) = \frac{1}{30,000}\left(1,800x - 3x^2\right) = \frac{1}{10,000}\left(600x - x^2\right)$$

$$R''(x) = \frac{1}{10,000}\left(600 - 2x\right) = \frac{1}{5,000}\left(300 - x\right)$$

Constructing a summary table for the second derivative:

Interval/Value	$R''(x)$	Behavior of $R(x)$
$(0,300)$	$(+)$	$R(x)$ is concave upward
300	0	$x = 200$ is an inflection point
$(300,900)$	$(-)$	$R(x)$ is concave downward

The function is concave upward on $(0,300)$, and concave downward on $(300,900)$. The point of diminishing returns is \$300 spent on advertising.

41. The hypercritical values are $x = 3$ and $x = 6$. Constructing a summary table for the second derivative:

Interval/Value	$f''(x)$	Behavior of $f(x)$
$(-\infty,3)$	$(-)$	$f(x)$ is concave downward
3	undefined	$x = 3$ is a vertical tangent
$(3,6)$	$(-)$	$f(x)$ is concave downward
6	0	$x = 6$ is an inflection point
$(6,\infty)$	$(+)$	$f(x)$ is concave upward

The function is concave downward on $(-\infty,3)$ and $(3,6)$, and concave upward on $(6,\infty)$.

43. The hypercritical values are $x = 6$ and $x = 9$. Constructing a summary table for the second derivative:

Interval/Value	$f''(x)$	Behavior of $f(x)$
$[0,6)$	$(+)$	$f(x)$ is concave upward
6	0	$x = 6$ is an inflection point
$(6,9)$	$(-)$	$f(x)$ is concave downward
9	0	$x = 9$ is an inflection point
$(9,\infty)$	$(+)$	$f(x)$ is concave upward

The function is concave upward on $[0,6)$ and $(9,\infty)$, and concave downward on $(6,9)$.

45. There are no hypercritical values. The function is concave upward on $(-\infty,\infty)$, and never concave downward.

47. In this case $f''(x) > 0$, so the curve is concave upward.

49. In this case $f''(x) < 0$, so the curve is concave downward.

51. In this case $f''(x) < 0$, so the curve is concave downward.

53. Yes, such a curve is possible.   **55.** Yes, such a curve is possible.

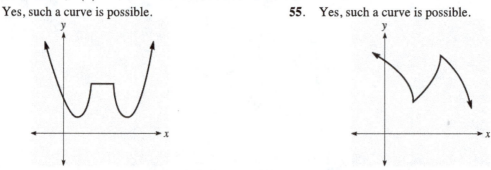

57. As the number of units produced increases towards 1,500, the average cost per unit decreases at a decreasing rate. At 1,500 units produced, the average cost reaches a minimum of $25. Then as the number of units produced increases beyond 1,500, the average cost per unit increases at an increasing rate.

59. As the increase in tuition increases from 0% to 70%, the average enrollment increases at an increasing rate towards 24,000 students. As the increase in tuition increases beyond 70%, the average enrollment increases at a decreasing rate towards 37,000 students, which is the limiting value.

61. Finding the first and second derivatives:

$$V'(t) = \frac{(t+3)(340) - (340t)(1)}{(t+3)^2} = \frac{340t + 1,020 - 340t}{(t+3)^2} = \frac{1,020}{(t+3)^2}$$

$$V''(t) = -2,040(t+3)^{-3} = -\frac{2,040}{(t+3)^3}$$

For $t > 0$, the function is increasing and concave down. Thus during the first 4 seconds, the speed is increasing at a decreasing rate. Based on the derivatives, you should reject the claim.

63. First write the function as $S(x) = \frac{1}{2}x(300-x)^{1/2}$. Finding the first and second derivatives:

$$S'(x) = \frac{1}{2}x \cdot \frac{1}{2}(300-x)^{-1/2}(-1) + \frac{1}{2}(300-x)^{1/2}$$

$$= -\frac{1}{4}x(300-x)^{-1/2} + \frac{1}{2}(300-x)^{1/2}$$

$$= \frac{1}{4}(300-x)^{-1/2}(-x + 2(300-x))$$

$$= \frac{1}{4}(300-x)^{-1/2}(600-3x)$$

$$= \frac{3(200-x)}{4(300-x)^{1/2}}$$

$$S''(x) = \frac{4(300-x)^{1/2} \cdot (-3) - 3(200-x) \cdot 2(300-x)^{-1/2}(-1)}{16(300-x)}$$

$$= \frac{-12(300-x)^{1/2} + 6(200-x)(300-x)^{-1/2}}{16(300-x)} \cdot \frac{(300-x)^{1/2}}{(300-x)^{1/2}}$$

$$= \frac{-6(300-x) + 3(200-x)}{8(300-x)^{3/2}}$$

$$= \frac{-1,800 + 6x + 600 - 3x}{8(300-x)^{3/2}}$$

$$= \frac{3x - 1,200}{8(300-x)^{3/2}}$$

$$= \frac{3(x-400)}{8(300-x)^{3/2}}$$

a. For $0 < x < 200$, $S'(x) > 0$ and $S''(x) < 0$, so the strength of the reaction is increasing at a decreasing rate. This supports the manufacturer's claim.

b. Since $S'(200) = 0$ and $S''(200) < 0$, the curve is concave down and $x = 200$ is a maximum point. The strength of the drug is at a maximum when $x = 200$, which supports the manufacturer's claim.

c. For $200 < x < 300$, $S'(x) < 0$ and $S''(x) < 0$, so the strength of the reaction is decreasing at an increasing rate. You should reject the manufacturer's claim.

65. Finding the first and second derivatives:

$$C'(x) = \frac{180x(2x) - (144 + x^2)(180)}{(180x)^2} = \frac{2x^2 - 144 - x^2}{180x^2} = \frac{x^2 - 144}{180x^2}$$

$$C''(x) = \frac{180x^2(2x) - (x^2 - 144)(360x)}{(180x^2)^2} = \frac{2x^3 - 2x^3 + 288x}{180x^4} = \frac{8}{5x^3}$$

For $x > 12$, $C'(x) > 0$ and $C''(x) > 0$, so once the manufacturer produces 1,200 systems, the costs will increase at an increasing rate.

67. Sketching a graph:

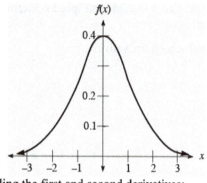

69. Sketching a graph:

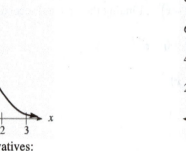

71. Finding the first and second derivatives:

$$f'(x) = 3x^2 - 8x$$
$$f''(x) = 6x - 8$$

Setting the second derivative equal to 0:

$$6x - 8 = 0$$
$$6x = 8$$
$$x = \frac{4}{3}$$

The hypercritical value occurs when $x = \frac{4}{3}$.

73. Finding the first and second derivatives:

$$L'(t) = 0.15(x - 4)^2 + 0.5$$
$$L''(t) = 0.3(x - 4)$$

For $4 < t < 7$, $L'(t) > 0$ and $L''(t) > 0$, so the trainee's skill level increases at an increasing rate.

75. Let w represent the width and $w + 20$ represent the length. Then the perimeter is given by:

$$P(w) = 2(w) + 2(w + 20) = 2w + 2w + 40 = 4w + 40$$

77. Setting the derivative equal to 0:

$$80 - 4w = 0$$
$$80 = 4w$$
$$w = 20$$

The critical value is $w = 20$.

79. Subtracting:

$$\left(400-10r\right)\left(20+2r\right)-200\left(20+2r\right)=\left(20+2r\right)\left(400-10r-200\right)$$
$$=\left(20+2r\right)\left(200-10r\right)$$
$$=4{,}000+400r-200r-20r^2$$
$$=4{,}000+200r-20r^2$$

81. Finding the derivative: $V'(h)=160-104h+12h^2$

83. Solving the equation:

$$\frac{0.21-0.03t}{\left(t+7\right)^3}=0$$
$$0.21=0.03t$$
$$t=\frac{0.21}{0.03}=7$$

3.3 Applications of the Derivative: Optimization

1. Finding the derivative: $C'(t)=\dfrac{\left(t^2+16\right)\left(100\right)-\left(100t\right)\left(2t\right)}{\left(t^2+16\right)^2}=\dfrac{100\left(t^2+16-2t^2\right)}{\left(t^2+16\right)^2}=\dfrac{100\left(16-t^2\right)}{\left(t^2+16\right)^2}=\dfrac{100\left(4+t\right)\left(4-t\right)}{\left(t^2+16\right)^2}$

The derivative is 0 when $t=4$ (note that $t\geq 0$). Evaluating the function:

$$C(0)=\frac{100(0)}{0^2+16}=0 \qquad\qquad C(4)=\frac{100(4)}{4^2+16}=\frac{400}{32}=\frac{25}{2}$$

The drug reaches its maximum concentration after 4 minutes.

3. First write the function as $C(r)=120r+4320r^{-1}$. Finding the derivative:

$$C'(r)=120-4320r^{-2}=120-\frac{4320}{r^2}=\frac{120\left(r^2-36\right)}{r^2}=\frac{120\left(r+6\right)\left(r-6\right)}{r^2}$$

The derivative is 0 when $r=6$ (note that $r>0$). Evaluating the function: $C(6)=120(6)+\dfrac{4{,}320}{6}=1{,}440$

The total cost is minimized if 6 inches of insulation are placed in the attic.

5. Finding the derivative: $R'(x)=-3x^2+126x+1{,}200$

Setting the derivative equal to 0:

$$-3x^2+126x+1{,}200=0$$
$$x^2-42x-400=0$$
$$\left(x-50\right)\left(x+8\right)=0$$

The derivative is 0 when $x=50$ (note that $x\geq 0$). Evaluating the function:

$$R(0)=-0^3+63(0)^2+1{,}200(0)=0$$
$$R(50)=-50^3+63(50)^2+1{,}200(50)=92{,}500$$

The revenue is maximum when $x=50$, which corresponds to a production level of 50,000 units.

7. The population reaches a maximum in approximately 1980.

9. The minimum occurs when each of the legs are 10 units in length.

11. The maximum profit occurs when $n=25$, which corresponds to 2,500 magazines sold.

13. Let x represent the width of each enclosure and h represent the height of each enclosure. Then:

$$4x+3h=1{,}200$$
$$3h=1{,}200-4x$$
$$h=400-\frac{4}{3}x$$

Substituting into the area: $A = xh = x\left(400 - \dfrac{4}{3}x\right) = 400x - \dfrac{4}{3}x^2$

Finding the derivative: $A'(x) = 400 - \dfrac{8}{3}x$

Setting the derivative equal to 0:

$$400 - \frac{8}{3}x = 0$$

$$400 = \frac{8}{3}x$$

$$x = 150$$

$$h = 400 - \frac{4}{3}(150) = 200$$

The farmer should build the enclosures 150 feet by 200 feet to maximize the area enclosed.

15. Let x represent the number of \$1 increases above \$50, thus $3x$ represents the reduction in rooms rented from 210. The revenue function is then given as:

$$R(x) = (\text{price})(\text{number of rooms})$$
$$= (50 + x)(210 - 3x)$$
$$= 10,500 - 150x + 210x - 3x^2$$
$$= 10,500 + 60x - 3x^2$$

Finding the derivative: $R'(x) = 60 - 6x$

Setting the derivative equal to 0:

$$60 - 6x = 0$$
$$60 = 6x$$
$$x = 10$$

The price they should charge is \$50 + \$10 = \$60 to maximize the daily revenue.

17. Let x represent the number of trees per acre above 30, thus $12x$ represents the reduction in yield from 480 peaches. The yield function is then given as:

$$Y(x) = (\text{yield per tree})(\text{number of trees})$$
$$= (480 - 12x)(30 + x)$$
$$= 14,400 + 480x - 360x - 12x^2$$
$$= 14,400 + 120x - 12x^2$$

Finding the derivative: $Y'(x) = 120 - 24x$

Setting the derivative equal to 0:

$$120 - 24x = 0$$
$$120 = 24x$$
$$x = 5$$

The number of trees should be 30 + 5 = 35, which will lead to a maximum yield per acre. Five additional trees per acre should be planted.

19. Let x represent the side of the square and l represent the length. The maximum girth is 100 inches, so:

$$4x + l = 100$$
$$l = 100 - 4x$$

The volume function is given as: $V(x) = x^2 l = x^2(100 - 4x) = 100x^2 - 4x^3$

Finding the derivative: $V'(x) = 200x - 12x^2$

Setting the derivative equal to 0:

$$200x - 12x^2 = 0$$
$$4x(50 - 3x) = 0$$
$$x = \frac{50}{3} = 16\frac{2}{3}$$
$$l = 100 - 4\left(\frac{50}{3}\right) = \frac{100}{3} = 33\frac{1}{3}$$

The length should be $33\frac{1}{3}$ inches and the square ends should be $16\frac{2}{3}$ inches.

21. Let x represent the number of $1 increases above $30, thus $10x$ represents the reduction in jackhammers rented from 500. The revenue function is then given as:

$$R(x) = (\text{price})(\text{number of jackhammers})$$
$$= (30 + x)(500 - 10x)$$
$$= 15{,}000 - 300x + 500x - 10x^2$$
$$= 15{,}000 + 200x - 10x^2$$

Finding the derivative: $R'(x) = 200 - 20x$

Setting the derivative equal to 0:

$$200 - 20x = 0$$
$$200 = 20x$$
$$x = 10$$

The price they should charge is $30 + $10 = $40 to maximize the revenue.

23. Finding the derivative: $f'(x) = \frac{5}{3}x^{2/3} - \frac{10}{3}x^{-1/3} = \frac{5}{3}x^{-1/3}(x - 2)$

The critical points are $x = 0$ and $x = 2$. Evaluating the function:

x	$f(x)$
0	3
2	−1.76
5	3

The absolute minimum value is approximately −1.76, which occurs when $x = 2$.

25. Using Wolfram Alpha, the minimum population is approximately 239,000, which occurs approximately 9 days after the insecticide is applied.

27. Finding the derivative:

$$L'(t) = \frac{\left(7.5 + 0.25t^{1.7}\right)(105) - (105t)\left(0.425t^{0.7}\right)}{\left(7.5 + 0.25t^{1.7}\right)^2} = \frac{105\left(7.5 + 0.25t^{1.7} - 0.425t^{1.7}\right)}{\left(7.5 + 0.25t^{1.7}\right)^2} = \frac{105\left(7.5 - 0.175t^{1.7}\right)}{\left(7.5 + 0.25t^{1.7}\right)^2}$$

Finding where the derivative is equal to 0:

$$7.5 - 0.175t^{1.7} = 0$$
$$7.5 = 0.175t^{1.7}$$
$$t^{1.7} = \frac{7.5}{0.175}$$
$$t = \left(\frac{7.5}{0.175}\right)^{1/1.7} \approx 9$$

The maximum interest level will be reached 9 minutes after the learning begins.

29. **a.** Finding the derivative: $C'(x) = -0.07x + 40$

  b. Evaluating the derivative: $C'(600) = -0.07(600) + 40 = -2$

31. Finding the derivative: $P'(x) = -0.06x^2 - 1.2x + 31.5$ **33.** Multiplying: $32 \cdot \dfrac{x}{2} = 16x$

35. Writing without a radical: $\dfrac{4{,}500}{\sqrt[3]{p^2}} = \dfrac{4{,}500}{p^{2/3}} = 4{,}500p^{-2/3}$ **37.** Simplifying: $500 \text{ units} \cdot \dfrac{\$27}{\text{unit}} = \$13{,}500$

3.4 Applications of the Derivative in Business and Economics

1. **a.** First find q: $q = 140 - 10(10) = 40$

  Now find $\dfrac{dq}{dp}$: $\dfrac{dq}{dp} = -10$

  Now find the point elasticity: $\varepsilon = -\dfrac{p}{q} \cdot \dfrac{dq}{dp} = -\dfrac{10}{40}(-10) = 2.5$

  b. Since $\varepsilon > 1$, the demand is elastic.

3. **a.** First find q: $q = \dfrac{5{,}000}{\sqrt[5]{32^2}} = 1{,}250$

  Writing $q = 5{,}000p^{-2/5}$, find the derivative: $\dfrac{dq}{dp} = -2{,}000p^{-7/5} = -\dfrac{2{,}000}{p^{7/5}}$

  Now evaluate $\dfrac{dq}{dp}$: $\dfrac{dq}{dp} = -\dfrac{2{,}000}{32^{7/5}} = -15.625$

  Now find the point elasticity: $\varepsilon = -\dfrac{p}{q} \cdot \dfrac{dq}{dp} = -\dfrac{32}{1{,}250}(-15.625) = 0.4$

  b. Since $\varepsilon < 1$, the demand is inelastic.

5. **a.** First find q: $q = \dfrac{1}{4}\left(400 - 10^2\right) = 75$

  Find the derivative: $\dfrac{dq}{dp} = \dfrac{1}{4}(-2p) = -\dfrac{1}{2}p$

  Now evaluate $\dfrac{dq}{dp}$: $\dfrac{dq}{dp} = -\dfrac{1}{2}(10) = -5$

  Now find the point elasticity: $\varepsilon = -\dfrac{p}{q} \cdot \dfrac{dq}{dp} = -\dfrac{10}{75}(-5) = \dfrac{2}{3}$

  b. Since $\varepsilon < 1$, the demand is inelastic.

7. **a.** First find q: $q = \dfrac{864}{\sqrt{36^3}} = 4$

Writing $q = 864 p^{-3/2}$, find the derivative: $\dfrac{dq}{dp} = -1{,}296 p^{-5/2} = -\dfrac{1{,}296}{p^{5/2}}$

Now evaluate $\dfrac{dq}{dp}$: $\dfrac{dq}{dp} = -\dfrac{1{,}296}{36^{5/2}} = -\dfrac{1}{6}$

Now find the point elasticity: $\varepsilon = -\dfrac{p}{q} \cdot \dfrac{dq}{dp} = -\dfrac{36}{4}\left(-\dfrac{1}{6}\right) = \dfrac{3}{2}$

b. Since $\varepsilon > 1$, the demand is elastic.

9. With a price of $8 and demand of 40,000, the revenue is ($8)(40,000) = $320,000.
A 1% increase in price results in a new price of $8 + 0.01($8) = $8.08.
The resulting effect on demand is 40,000 − 0.005(40,000) = 39,800.
The resulting revenue is ($8.08)(39,800) = $321,584.
Thus the revenue increased by $1,584.

11. With a price of $20 and demand of 150,000, the revenue is ($20)(150,000) = $3,000,000.
A 1% increase in price results in a new price of $20 + 0.01($20) = $20.20.
The resulting effect on demand is 150,000 − 0.016(150,000) = 147,600.
The resulting revenue is ($20.20)(147,600) = $2,981,520.
Thus the revenue decreased by $18,480.

13. $C(250) = 12{,}000$ means the cost of producing 250 units is $12,000. $C'(250) = -85$ means the cost will drop by $85 to produce the next (251st) unit.

15. $P(55) = 16$ means the profit is $16,000 from the sale of units priced at $55. $P'(55) = -4$ means the profit will decrease by $4,000 if the units are priced $1 higher ($56).

17. $C(200) = 1{,}500$ means the inventory cost is $1,500 if the lot size is 200 units. $C'(200) = 300$ means the inventory cost will increase by $300 if the lot size is increased by 1 unit (to 201 units).

19. Since $\varepsilon > 1$, the demand is elastic. The revenue will decrease if the price of the product is increased.

21. Since $\varepsilon = 1$, the demand is unit elastic. The revenue will stay the same if the price of the product is increased.

23. First find the derivative: $C'(x) = 3x + 45$
Finding when the marginal cost is $210:
$$3x + 45 = 210$$
$$3x = 165$$
$$x = 55$$
The most units the manufacturer can make is 55 units. At that level, the total cost is:
$$C(55) = \frac{3}{2}(55)^2 + 45(55) + 720 = \$7{,}732.50$$

25. First find the derivative: $P'(x) = -0.5x + 80$
Evaluating the marginal profit:
$$P'(100) = -0.5(100) + 80 = 30$$
$$P'(260) = -0.5(260) + 80 = -50$$
The marginal profit is $30 at the $100,000 level of advertising, and it is −$50 at the $260,000 level of advertising.
Finding when the marginal profit is 0:
$$-0.5x + 80 = 0$$
$$80 = 0.5x$$
$$x = 160$$
To maximize profit, the company should spend $160,000 on advertising.

27. Let x represent the number of 0.25 decreases in price below 75, so the revenue is given by:
$$R(x) = (\text{price})(\text{quantity})$$
$$= (75 - 0.25x)(3{,}000 + 20x)$$
$$= 225{,}000 - 750x + 1{,}500x - 5x^2$$
$$= 225{,}000 + 750x - 5x^2$$

The cost function is given by:
$$C(x) = 350 + 4.20(3{,}000 + 20x) = 350 + 12{,}600 + 84x = 12{,}950 + 84x$$

Thus the profit is given by:
$$P(x) = R(x) - C(x)$$
$$= (225{,}000 + 750x - 5x^2) - (12{,}950 + 84x)$$
$$= 212{,}050 + 666x - 5x^2$$

Finding the derivative: $P'(x) = 666 - 10x = 10(66.6 - x)$

The profit will have a maximum value when $x = 66.6$. The number of units sold is: $3{,}000 + 20(66.6) = 4{,}332$

The maximum profit is: $P(66.6) = 212{,}050 + 666(66.6) - 5(66.6)^2 = \$234{,}227.80$

29. Let x represent the lot size. Finding the carrying cost and reordering cost:
$$H(x) = 10 \cdot \frac{x}{2} = 5x$$
$$R(x) = 15 \cdot \frac{300}{x} = \frac{4{,}500}{x}$$

Thus the total inventory cost is given by: $C(x) = 5x + \dfrac{4{,}500}{x} = 5x + 4{,}500x^{-1}$

Finding the derivative: $C'(x) = 5 - 4{,}500x^{-2} = 5 - \dfrac{4{,}500}{x^2}$

Setting the derivative equal to 0:
$$5 - \frac{4{,}500}{x^2} = 0$$
$$5x^2 = 4{,}500$$
$$x^2 = 900$$
$$x = 30$$

Note that only the positive root was chosen. The lot size is 30 fuel and water separators, and the number of orders is $\dfrac{300}{30} = 10$ orders. The minimum inventory cost is: $C(30) = 5(30) + \dfrac{4{,}500}{30} = \$300$

31. Let x represent the lot size. Finding the carrying cost and reordering cost:
$$H(x) = 4 \cdot \frac{x}{2} = 2x$$
$$R(x) = 200 \cdot \frac{40{,}000}{x} = \frac{8{,}000{,}000}{x}$$

Thus the total inventory cost is given by: $C(x) = 2x + \dfrac{8{,}000{,}000}{x} = 2x + 8{,}000{,}000x^{-1}$

Finding the derivative: $C'(x) = 2 - 8{,}000{,}000x^{-2} = 2 - \dfrac{8{,}000{,}000}{x^2}$

Setting the derivative equal to 0:

$$2 - \frac{8{,}000{,}000}{x^2} = 0$$

$$2x^2 = 8{,}000{,}000$$

$$x^2 = 4{,}000{,}000$$

$$x = 2{,}000$$

Note that only the positive root was chosen. The lot size is 2,000 reams of paper, and the number of orders is $\frac{40{,}000}{2{,}000} = 20$ orders. The minimum inventory cost is: $C(600) = 2(2{,}000) + \frac{8{,}000{,}000}{2{,}000} = \$8{,}000$

Nick's conclusion is accurate.

33. Using the EOQ formula: $x = \sqrt{\dfrac{2 \cdot 15 \cdot 300}{10}} = 30$. This matches the lot size found in Problem 29.

35. First find q: $q = \dfrac{14{,}625}{\sqrt[4]{4^{11}}} \approx 323.17$

Writing $q = 14{,}625\, p^{-11/4}$, find the derivative: $\dfrac{dq}{dp} = -40{,}218.75\, p^{-15/4} = -\dfrac{40{,}218.75}{p^{15/4}}$

Now evaluate $\dfrac{dq}{dp}$: $\dfrac{dq}{dp} = -\dfrac{40{,}218.75}{4^{15/4}} = -222.18$

Now find the point elasticity: $\varepsilon = -\dfrac{p}{q} \cdot \dfrac{dq}{dp} = -\dfrac{4}{323.17}(-222.18) \approx 2.75$

Since $\varepsilon > 1$, the demand is elastic. If the price is raised by 1%, the demand will decrease by 2.75% and the revenue will decrease.

37. With a price of $160 and demand of 35,000, the revenue is ($160)(35,000) = $5,600,000.
A 1% increase in price results in a new price of $160 + 0.01($160) = $161.60.
The resulting effect on demand is 35,000 − 0.088(35,000) = 31,920.
The resulting revenue is ($161.60)(31,920) = $5,158,272.
Thus the revenue decreased by $441,728.

39. First find q: $q = 4{,}500(30) - 2(30)^3 = 81{,}000$

Finding the derivative: $\dfrac{dq}{dp} = 4{,}500 - 6p^2$

Now evaluate $\dfrac{dq}{dp}$: $\dfrac{dq}{dp} = 4{,}500 - 6(30)^2 = -900$

Now find the point elasticity: $\varepsilon = -\dfrac{p}{q} \cdot \dfrac{dq}{dp} = -\dfrac{30}{81{,}000}(-900) = \dfrac{1}{3}$

Since $\varepsilon < 1$, the demand is inelastic.

41. First find q: $q = 2{,}160(8)^{1.1} \approx 21{,}274.18$

Finding the derivative: $\dfrac{dq}{dp} = 2{,}376\, p^{0.1}$

Now evaluate $\dfrac{dq}{dp}$: $\dfrac{dq}{dp} = 2{,}376(8)^{0.1} \approx 2{,}925.20$

Note that $\dfrac{dq}{dp} > 0$, so we don't need the negative sign in the point elasticity formula.

Now find the point elasticity: $\varepsilon = \dfrac{p}{q} \cdot \dfrac{dq}{dp} = \dfrac{8}{21{,}274.18}(2{,}925.20) \approx 1.1$

Since $\varepsilon > 1$, the demand is elastic.

43. With a price of $50 and demand of 2,000, the revenue is ($50)(2,000) = $100,000.
A 1% increase in price results in a new price of $50 + 0.01($50) = $50.50.
The resulting effect on demand is 2,000 − 0.0073(2,000) = 1,985.4.
The resulting revenue is ($50.50)(1,985.4) = $100,262.70.
Thus the revenue will increase by $262.70.

45. Computing:

$$2^{-3} = \frac{1}{2^3} = \frac{1}{8}$$

$$2^3 = 8$$

47. Computing the value: $5e^{0.06931(120)} = 5e^{8.3172} \approx 20,468$

49. Computing the value: $2,000\left(1 + \frac{0.08}{4}\right)^{4(15)} = 2,000(1.02)^{60} \approx 6,562.06$

Chapter 3 Test

1. The critical values are 1, 8, and 14. The function is decreasing on $(-\infty, 1)$ and $(8, 14)$, and the function is increasing on $(1, 8)$ and $(14, \infty)$.

2. The hypercritical values are 8 and 21. The function is concave down on $(21, \infty)$, and the function is concave up on $(-\infty, 8)$ and $(8, 21)$.

3. The relative and absolute minimum occurs at $x = 2$, and the relative and absolute maximum occurs at $x = 5$.

4. A curve that is increasing at a decreasing rate is concave downward.

5. A curve that is decreasing at an increasing rate is concave downward.

6. Constructing a sign chart:

$f'(x)$	(−)		(+)
$f''(x)$	(+)	6	(−)

7. The maximum volume occurs when $x = 4$. The corners should be cut 4 units, and the maximum volume is approximately 1,150 cubic units.

8. Constructing a summary table:

Interval/Value	$f'(x)$	$f''(x)$	Behavior of $f(x)$
$(-\infty, -5)$	(+)	(+)	$f(x)$ is increasing and concave up
−5	(+)	0	$x = -5$ is a point of inflection
$(-5, 9)$	(+)	(−)	$f(x)$ is increasing and concave down
9	0	(−)	$x = 9$ is a relative maximum
$(9, 25)$	(−)	(−)	$f(x)$ is decreasing and concave down
25	(−)	0	$x = 25$ is a point of inflection
$(25, \infty)$	(−)	(+)	$f(x)$ is decreasing and concave up

9. The amount of money in the account increases at an increasing rate.

10. For the first 5 hours, the rate of production of antibodies increases at a decreasing rate. After 5 hours, the rate of production reaches a maximum value. From 5 to 6 hours, the rate of production decreases at an increasing rate. From 6 hours on, the rate of production decreases at a decreasing rate.

11. Finding the derivative: $f'(x) = 24x - 17$

Setting the derivative equal to 0:

$$24x - 17 = 0$$
$$24x = 17$$
$$x = \frac{17}{24}$$

12. Finding the first and second derivatives:

$$f'(x) = \frac{1}{3}x^3 - \frac{7}{2}x^2 + 5$$
$$f''(x) = x^2 - 7x$$

Setting the second derivative equal to 0:

$$x^2 - 7x = 0$$
$$x(x - 7) = 0$$
$$x = 0, 7$$

The hypercritical values are $x = 0$ and $x = 7$.

13. Since $f''(0) = -8 < 0$, $x = 0$ produces a relative maximum. Since $f''(-2) = 8 > 0$, $x = -2$ produces a relative minimum. Since $f''(2) = 8 > 0$, $x = 2$ produces a relative minimum.

14. Finding the derivative: $f'(x) = 3x^2 - 10x - 36$

The critical points occur when $f'(x) = 0$. Using the quadratic formula:

$$x = \frac{10 \pm \sqrt{(-10)^2 - 4(3)(-36)}}{2(3)} = \frac{10 \pm \sqrt{532}}{6} = \frac{10 \pm 2\sqrt{133}}{6} = \frac{5 \pm \sqrt{133}}{3}$$

The function is increasing on $\left(-\infty, \frac{5 - \sqrt{133}}{3}\right)$ and $\left(\frac{5 + \sqrt{133}}{3}, \infty\right)$, and the function is decreasing on $\left(\frac{5 - \sqrt{133}}{3}, \frac{5 + \sqrt{133}}{3}\right)$.

15. Finding the first and second derivatives:

$$f'(x) = 4x^3 - 24x^2 - 64x$$
$$f''(x) = 12x^2 - 48x - 64$$

Find the critical values by setting the first derivative equal to 0:

$$4x^3 - 24x^2 - 64x = 0$$
$$x^3 - 6x^2 - 16x = 0$$
$$x(x^2 - 6x - 16) = 0$$
$$x(x - 8)(x + 2) = 0$$
$$x = -2, 0, 8$$

Evaluating the second derivative at these critical values:

$$f''(-2) = 12(-2)^2 - 48(-2) - 64 = 80 > 0 \qquad \text{relative minimum}$$
$$f''(0) = 12(0)^2 - 48(0) - 64 = -64 < 0 \qquad \text{relative maximum}$$
$$f''(8) = 12(8)^2 - 48(8) - 64 = 320 > 0 \qquad \text{relative minimum}$$

Finally, evaluating the function:
$$f(-2) = (-2)^4 - 8(-2)^3 - 32(-2)^2 + 20 = -28$$
$$f(0) = (0)^4 - 8(0)^3 - 32(0)^2 + 20 = 20$$
$$f(8) = (8)^4 - 8(8)^3 - 32(8)^2 + 20 = -2,028$$
The function has a relative maximum at $(0,20)$, and relative minima at $(-2,-28)$ and $(8,-2028)$.

16. Write the function as $f(x) = 5(x-4)^{-1}$. Finding the derivative: $f'(x) = -5(x-4)^{-2} = \dfrac{-5}{(x-4)^2}$

Since the function is decreasing everywhere, it will have an absolute maximum when $x = 5$ (the left endpoint of the interval). Since $f(5) = 5$, the absolute maximum is at $(5,5)$.

17. Finding the first and second derivatives:
$$f'(x) = 3(2x+2)^2(2) = 6(2x+2)^2$$
$$f''(x) = 12(2x+2)(2) = 48(x+1)$$
The function is concave downward on $(-\infty,-1)$ and concave upward on $(-1,\infty)$.

18. Let x represent the number of $0.50 reductions in price below $100, so the revenue is given by:
$$R(x) = (\text{price})(\text{quantity})$$
$$= (100 - 0.5x)(40 + 20x)$$
$$= 4,000 - 20x + 2,000x - 10x^2$$
$$= 4,000 + 1,800x - 10x^2$$
The cost function is given by:
$$C(x) = 60(40 + 20x) = 2,400 + 1,200x$$
Thus the profit is given by:
$$P(x) = R(x) - C(x)$$
$$= (4,000 + 1,800x - 10x^2) - (2,400 + 1,200x)$$
$$= 1,600 + 600x - 10x^2$$
Finding the derivative: $P'(x) = 600 - 20x = 20(30 - x)$

So to maximize profit the company should make 30 reductions in price. The new price would then be $100 − 30($0.50) = $85, and the quantity is 40 + 20(30) = 640. This will result in a profit of:
$$P(30) = 1,600 + 600(30) - 10(30)^2 = \$10,600$$
The company should sell 640 filters each week, priced at $85, to maximize their profit at $10,600.

19. $C(370) = 5,000$ means the inventory cost is $5,000 if the lot size is 370 units. $C'(370) = 200$ means the inventory cost will increase by $200 if the lot size is increased by 1 unit (to 371 units).

20. With a price of $40 and demand of 1,000, the revenue is ($40)(1,000) = $40,000.
A 1% increase in price results in a new price of $40 + 0.01($40) = $40.40.
The resulting effect on demand is 1,000 − 0.0074(1,000) = 992.6.
The resulting revenue is ($40.40)(992.6) = $40,101.04.
Thus the revenue will increase by $101.04.

21. **a.** Finding the actual costs:
$$C(401) - C(400) \approx 13,522.60 - 13,520 = \$2.60$$
$$C(2,001) - C(2,000) \approx 9,993.00 - 10,000 = -\$7.00$$

b. First find the derivative: $C'(x) = -0.006x + 5$
Finding the marginal costs:
$$C'(400) = -0.006(400) + 5 = \$2.60$$
$$C'(2,000) = -0.006(2,000) + 5 = -\$7.00$$

22. The sign of $U'(x)$ is positive, as the total utility is increasing. The sign of $U''(x)$ is negative, as the marginal utility of each additional item is decreasing.

23. Let x and y represent the dimensions of the inside portion containing photographs, therefore:

$$xy = 24$$

$$y = \frac{24}{x}$$

The entire page has dimensions $x + 2$ and $y + 3$ (remember there are margins on both sides), so the area function is given by:

$$A(x) = (x+2)\left(\frac{24}{x}+3\right) = 24 + \frac{48}{x} + 3x + 6 = 30 + 3x + 48x^{-1}$$

Finding the derivative: $A'(x) = 3 - 48x^{-2} = 3 - \frac{48}{x^2}$

Setting the derivative equal to 0:

$$3 - \frac{48}{x^2} = 0$$

$$3x^2 - 48 = 0$$

$$3x^2 = 48$$

$$x^2 = 16$$

$$x = 4$$

So the dimensions of the page are $4 + 2 = 6$ by $\frac{24}{4} + 3 = 9$. The page is 6 inches by 9 inches.

24. Using the EOQ formula: $x = \sqrt{\dfrac{2 \cdot 500 \cdot 15{,}000}{15}} = 1{,}000$. The lot size is 1,000 units that will minimize inventory costs.

Chapter 4
The Natural Exponential and Logarithmic Functions

4.1 The Exponential Functions

1. Evaluating the function: $f(3) = 10^3 = 1,000$
   This is an exponential growth function.

3. Evaluating the function: $f(6) = (0.99)^{240} \approx 0.090$
   This is an exponential decay function.

5. Evaluating the function: $f(5) = (1.35)^{19} \approx 299.462$
   This is an exponential growth function.

7. Evaluating the function: $f(3) = \left(\dfrac{1}{2}\right)^4 = \dfrac{1}{16} \approx 0.063$

   This is an exponential decay function.

9. Evaluating the function: $f(2) = e^{-2} \approx 0.135$
   This is an exponential decay function.

11. Evaluating the function: $f(105) = 3,560e^{0.06(105)} = 3,560e^{6.3} \approx 1,938,676$
    This is an exponential growth function.

13. Evaluating the function: $f(0.10) = 1 - e^{-0.3(0.10)} = 1 - e^{-0.03} \approx 0.030$

15. Evaluating the function: $f(3) = 100 - 60\left(\dfrac{1}{4}\right)^{0.95(3)} = 100 - 60\left(\dfrac{1}{4}\right)^{2.85} \approx 98.846$

17. a. This graph is D, since it has the smallest exponent.
    b. This graph is B, since it has the second largest exponent.
    c. This graph is C, since it has the second smallest exponent.
    d. This graph is A, since it has the largest exponent.

19. This function is algebraic.   21. This function is exponential.
23. This function is algebraic.   25. This function is exponential.
27. This function is exponential.   29. This function is algebraic.
31. If the base $b > 1$ the function is called an exponential growth function, and if $0 < b < 1$ the function is called an exponential decay function.

33. Finding the value when $t = 5$ and $t = 10$:

    $$V(5) = 4,500 \cdot 2^{5/5} = 4,500 \cdot 2 = \$9,000 \qquad V(10) = 4,500 \cdot 2^{10/5} = 4,500 \cdot 4 = \$18,000$$

    It appears the owner has overstated its value by \$1,000 after 5 years and by \$7,000 after 10 years.

35. a. Evaluating when $t = 60$: $A(60) = 2e^{0.13863(60)} = 2e^{8.3178} \approx 8,192$
       There are 8,192 bacteria present after 1 hour.
    b. Evaluating when $t = 120$: $A(120) = 2e^{0.13863(120)} = 2e^{16.6356} \approx 33,556,703$
       There are 33,556,703 bacteria present after 2 hours.

37. **a.** Evaluating: $N(0) = 15{,}000 \cdot 3^{-0.05(0)} = 15{,}000 \cdot 3^0 = 15{,}000$

b. Evaluating: $N(1) = 15{,}000 \cdot 3^{-0.05(1)} = 15{,}000 \cdot 3^{-0.05} \approx 14{,}198$

c. Evaluating: $N(5) = 15{,}000 \cdot 3^{-0.05(5)} = 15{,}000 \cdot 3^{-0.25} \approx 11{,}398$

d. Evaluating: $N(24) = 15{,}000 \cdot 3^{-0.05(24)} = 15{,}000 \cdot 3^{-1.2} \approx 4{,}014$

39. **a.** Substituting $P = 500, r = 0.11, t = 5,$ and $n = 1$: $A(5) = 500\left(1 + \dfrac{0.11}{1}\right)^{1 \cdot 5} \approx \$842.53$

b. Substituting $P = 500, r = 0.11, t = 5,$ and $n = 4$: $A(5) = 500\left(1 + \dfrac{0.11}{4}\right)^{4 \cdot 5} \approx \$860.21$

c. Substituting $P = 500, r = 0.11, t = 5,$ and $n = 12$: $A(5) = 500\left(1 + \dfrac{0.11}{12}\right)^{12 \cdot 5} \approx \$864.46$

d. Substituting $P = 500, r = 0.11, t = 5,$ and $n = 365$: $A(5) = 500\left(1 + \dfrac{0.11}{365}\right)^{365 \cdot 5} \approx \$866.55$

41. **a.** Substituting $P = 1{,}600, r = 0.09, t = 5,$ and $n = 4$: $A(5) = 1{,}600\left(1 + \dfrac{0.09}{4}\right)^{4 \cdot 5} \approx \$2{,}496.81$

b. Substituting $P = 1{,}600, r = 0.09, t = 5,$ and $n = 12$: $A(5) = 1{,}600\left(1 + \dfrac{0.09}{12}\right)^{12 \cdot 5} \approx \$2{,}505.09$

c. Substituting $P = 1{,}600, r = 0.09, t = 5,$ and $n = 365$: $A(5) = 1{,}600\left(1 + \dfrac{0.09}{365}\right)^{365 \cdot 5} \approx \$2{,}509.16$

d. Substituting $P = 1{,}600, r = 0.09,$ and $t = 5$: $A(5) = 1{,}600e^{0.09 \cdot 5} \approx \$2{,}509.30$

e. The values are very close.

43. **a.** Substituting $t = 0$: $C(0) = \dfrac{240}{1 + e^{-0.2(0)}} = 120$ customers

b. The maximum attainable level is 240 customers.

45. **a.** Substituting $i = 400$: $T(400) = \dfrac{0.9}{1 + 20e^{-400/400}} \approx 0.1$ telephones

b. Substituting $i = 4{,}000$: $T(4{,}000) = \dfrac{0.9}{1 + 20e^{-4000/400}} \approx 0.9$ telephones

47. **a.** Substituting $t = 15$: $A(15) = \dfrac{1}{0.20}\left(2 - 35e^{-0.20(15)}\right) \approx 1.3$ mg

b. Substituting $t = 60$: $A(60) = \dfrac{1}{0.20}\left(2 - 35e^{-0.20(60)}\right) \approx 10.0$ mg

49. **a.** Substituting $t = 0$: $N(0) = \dfrac{45}{1 + 38e^{-0.095(0)}} \approx 1{,}154$ people

b. Substituting $t = 3$: $N(3) = \dfrac{45}{1 + 38e^{-0.095(3)}} \approx 1{,}521$ people

c. Substituting $t = 10$: $N(10) = \dfrac{45}{1 + 38e^{-0.095(10)}} \approx 2{,}867$ people

d. Substituting $t = 26$: $N(26) = \dfrac{45}{1 + 38e^{-0.095(26)}} \approx 10{,}678$ people

e. In the long run 45,000 people will have contracted the disease.

51. Sketching the graph:

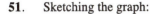

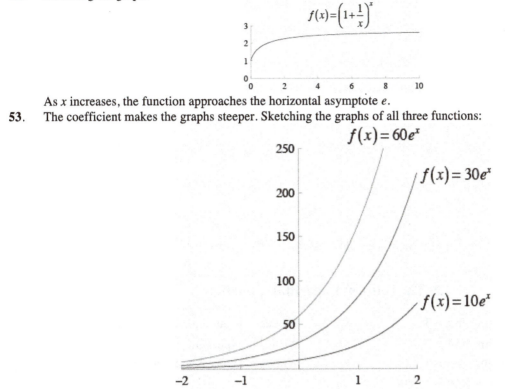

As x increases, the function approaches the horizontal asymptote e.

53. The coefficient makes the graphs steeper. Sketching the graphs of all three functions:

55. The function is increasing rapidly at first, then leveling off to the horizontal asymptote 600, which is the numerator of the function:

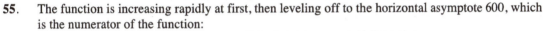

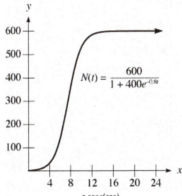

57. a. Substituting $t = 200$: $P(200) = 0.0150 + 0.985e^{-0.0086(200)} \approx 0.191 = 19.1\%$

    b. Substituting $t = 1,000$: $P(1,000) = 0.0150 + 0.985e^{-0.0086(1,000)} \approx 0.015 = 1.5\%$

59. A logarithm is an exponent.

61. The logarithmic form is: $\ln y = 5x$

63. Sketching the graph:

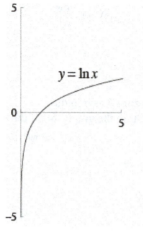

$y = \ln x$

65. Expanding the logarithm: $\ln\left(\dfrac{x^3 y^2}{z^4}\right) = \ln x^3 + \ln y^2 - \ln z^4 = 3\ln x + 2\ln y - 4\ln z$

4.2 The Natural Logarithm Function

1. Writing in logarithmic form: $\ln m = 3n$

3. Writing in logarithmic form: $\ln y = 3x - 5$

5. Writing in logarithmic form: $\ln 4 = 3k + 1$

7. Writing in exponential form: $4x = e^6$

9. Writing in exponential form: $y + 1 = e^{x+3}$

11. Writing in exponential form: $3x + 4 = e^{-7.18}$

13. Expanding the logarithm: $\ln 5x = \ln 5 + \ln x$

15. Expanding the logarithm: $\ln\dfrac{5}{y^2} = \ln 5 - \ln\left(y^2\right) = \ln 5 - 2\ln y$

17. Expanding the logarithm: $\ln\dfrac{10x}{x-6} = \ln 10 + \ln x - \ln(x-6)$

19. Writing as a single logarithm: $\ln 8 + \ln x = \ln(8x)$

21. Writing as a single logarithm: $\ln x - 3\ln y - 5\ln z = \ln x - \ln\left(y^3\right) - \ln\left(z^5\right) = \ln\left(\dfrac{x}{y^3 z^5}\right)$

23. Writing as a single logarithm: $\ln(x+2) - \ln(x+7) = \ln\left(\dfrac{x+2}{x+7}\right)$

25. Solving the equation:

$$e^{3x+7} = 12$$
$$3x + 7 = \ln 12$$
$$3x = \ln 12 - 7$$
$$x = \dfrac{\ln 12 - 7}{3} \approx -1.5050$$

27. Solving the equation:

$$31e^{8x-6} = 178$$
$$e^{8x-6} = \dfrac{178}{31}$$
$$8x - 6 = \ln\left(\dfrac{178}{31}\right)$$
$$8x = \ln\left(\dfrac{178}{31}\right) + 6$$
$$x = \dfrac{\ln\left(\dfrac{178}{31}\right) + 6}{8} \approx 0.9685$$

29. Solving the equation:

$$e^{-0.03t} = 0.4724$$

$$-0.03t = \ln 0.4724$$

$$t = \frac{\ln 0.4724}{-0.03} \approx 24.9976$$

31. Evaluating: $A(20) = 5e^{-0.076(20)} \approx 1.09 \text{ cm}^2$

33. Evaluating: $P(3.5) = 14.7e^{-0.21(3.5)} \approx 7.05 \text{ lb/in}^2$

35. Evaluating: $P(6) = 25.5e^{0.012(6)} \approx 27.4$ million

37. Evaluating: $N(21) = 125 - 125e^{-0.08(21)} \approx 102$ tasks

39. Evaluating: $N(15) = 1500e^{0.03(15)} \approx 2,352$ people

41. The first model (with $k = -0.0231$) will be the model with the 30 hour half-life, since the decay rate (2.31%) is greater than the second model (with $k = -0.0173$) with a decay rate of 1.73%.

43. Evaluating: $MV(15) = 240,000e^{0.2(15)^{2/5}} \approx \$433,319.06$

45. Evaluating: $T(7) = 68 + (98.6 - 68)e^{-0.133543(7)} \approx 81\Upsilon$

47. Solving the equation:

$$6.5 = 14.7e^{-0.21h}$$

$$e^{-0.21h} = \frac{6.5}{14.7}$$

$$-0.21h = \ln\left(\frac{6.5}{14.7}\right)$$

$$h = \frac{\ln\left(\dfrac{6.5}{14.7}\right)}{-0.21} \approx 3.9 \text{ miles}$$

49. Solving the equation:

$$119,175 = 36,000e^{0.015t}$$

$$e^{0.015t} = \frac{119,175}{36,000}$$

$$0.015t = \ln\left(\frac{119,175}{36,000}\right)$$

$$t = \frac{\ln\left(\dfrac{119,175}{36,000}\right)}{0.015} \approx 79.8 \text{ years}$$

51. Solving the equation:

$$114 = 125 - 125e^{-0.08t}$$

$$-11 = -125e^{-0.08t}$$

$$e^{-0.08t} = \frac{11}{125}$$

$$-0.08t = \ln\left(\frac{11}{125}\right)$$

$$t = \frac{\ln\left(\dfrac{11}{125}\right)}{-0.08} \approx 30 \text{ days}$$

53. Solving the equation:

$$55 = \frac{1}{0.52}\left(50 - 50e^{-0.12t}\right)$$

$$28.6 = 50 - 50e^{-0.12t}$$

$$-50e^{-0.12t} = -21.4$$

$$e^{-0.12t} = \frac{21.4}{50}$$

$$-0.12t = \ln\left(\frac{21.4}{50}\right)$$

$$t = \frac{\ln\left(\dfrac{21.4}{50}\right)}{-0.12} \approx 7.07 \text{ minutes}$$

55. Solving the equation:

$$0.50A_0 = A_0e^{-0.0248t}$$

$$e^{-0.0248t} = 0.5$$

$$-0.0248t = \ln 0.5$$

$$t = \frac{\ln 0.5}{-0.0248} \approx 27.9 \text{ years}$$

57.   **a.**   Evaluating: $A(6) = 10e^{-0.0231(6)} \approx 8.7$ mg

       b.   Solving the equation:

$$6.25 = 10e^{-0.0231t}$$

$$e^{-0.0231t} = 0.625$$

$$-0.0231t = \ln 0.625$$

$$t = \frac{\ln 0.625}{-0.0231} \approx 20 \text{ hours}$$

59.   **a.**   Solving the equation:

$$2A_0 = A_0 e^{0.06t}$$

$$e^{0.06t} = 2$$

$$0.06t = \ln 2$$

$$t = \frac{\ln 2}{0.06} \approx 11.55 \text{ years}$$

       b.   Solving the equation:

$$2A_0 = A_0 e^{0.08t}$$

$$e^{0.08t} = 2$$

$$0.08t = \ln 2$$

$$t = \frac{\ln 2}{0.08} \approx 8.66 \text{ years}$$

       c.   Solving the equation:

$$2A_0 = A_0 e^{0.10t}$$

$$e^{0.10t} = 2$$

$$0.10t = \ln 2$$

$$t = \frac{\ln 2}{0.10} \approx 6.93 \text{ years}$$

61.   Solving the equation:

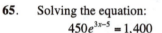

$$30{,}000 = A_0 e^{0.08(10)}$$

$$A_0 = \frac{30{,}000}{e^{0.8}} \approx \$13{,}479.87$$

63.   Solving the equation:

$$10{,}000 = 2{,}300 e^{0.195t}$$

$$e^{0.195t} = \frac{100}{23}$$

$$0.195t = \ln \frac{100}{23}$$

$$t = \frac{\ln \frac{100}{23}}{0.195} \approx 7.5 \text{ years}$$

65.   Solving the equation:

$$450e^{3x-5} = 1{,}400$$

$$e^{3x-5} = \frac{28}{9}$$

$$3x - 5 = \ln \frac{28}{9}$$

$$3x = 5 + \ln \frac{28}{9}$$

$$x = \frac{5 + \ln \frac{28}{9}}{3} \approx 2.04$$

67.   Solving the equation:

$$0.05(15{,}000) = 15{,}000 e^{-0.32t}$$

$$e^{-0.32t} = 0.05$$

$$-0.32t = \ln 0.05$$

$$x = \frac{\ln 0.05}{-0.32} \approx 9.4 \text{ years}$$

69. Solving the equation:

$$480{,}000 = 240{,}000e^{0.2t^{2/5}}$$

$$e^{0.2t^{2/5}} = 2$$

$$0.2t^{2/5} = \ln 2$$

$$t^{2/5} = \frac{\ln 2}{0.2}$$

$$t = \left(\frac{\ln 2}{0.2}\right)^{5/2} \approx 22.4 \text{ years}$$

Graphing the function:

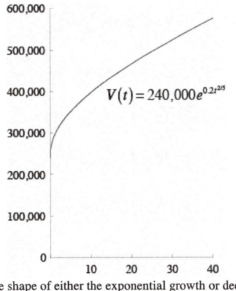

No, this graph does not exhibit the shape of either the exponential growth or decay function.

71. Simplifying: $e^1 = e$

73. Simplifying: $3\ln e + 1 = 3\ln e^1 + 1 = 3 + 1 = 4$

75. Differentiating: $\dfrac{dy}{dx} = 8 \cdot 2x = 16x$

77. Differentiating: $\dfrac{dy}{dx} = 10x + 2$

79. Approximating: $7e^2 \approx 51.723$

81. Approximating: $\ln 1.02 \approx 0.020$

83. Expanding the logarithm: $\ln\left(xy^2\right) = \ln x + \ln\left(y^2\right) = \ln x + 2\ln y$

4.3 Differentiating the Natural Logarithm Function

1. Finding the derivative: $f'(x) = \dfrac{1}{2x-7} \cdot 2 = \dfrac{2}{2x-7}$

3. Finding the derivative: $f'(x) = \dfrac{8}{4x^2} \cdot 8x = \dfrac{16}{x}$

5. Finding the derivative: $f'(x) = x^2 \cdot \dfrac{1}{3x} \cdot 3 + 2x\ln(3x) = x + 2x\ln(3x)$

7. Finding the derivative: $f'(x) = \dfrac{1}{(2x+8)^5} \cdot 5(2x+8)^4 \cdot 2 = \dfrac{10}{2x+8} = \dfrac{5}{x+4}$

9. Finding the derivative: $f'(w) = 3\left[\ln(w+4)\right]^2 \cdot \dfrac{1}{w+4} = \dfrac{3\left[\ln(w+4)\right]^2}{w+4}$

11. Finding the derivative: $f'(t) = \dfrac{1}{\sqrt{3t+5}} \cdot \dfrac{1}{2}(3t+5)^{-1/2} \cdot 3 = \dfrac{3}{2(3t+5)}$

13. Finding the derivative: $f'(u) = \dfrac{1}{\ln u} \cdot \dfrac{1}{u} = \dfrac{1}{u \ln u}$

15. Finding the derivative (using the product rule):

$$f'(t) = t^2 \cdot 2\left[\ln\left(t^2\right)\right] \cdot \dfrac{1}{t^2} \cdot 2t + 2t\left[\ln\left(t^2\right)\right]^2 = 4t\left[\ln\left(t^2\right)\right] + 2t\left[\ln\left(t^2\right)\right]^2 = 2t\ln\left(t^2\right)\left(2 + \ln\left(t^2\right)\right)$$

17. Finding the derivative: $f'(x) = \dfrac{1}{x/a} \cdot \dfrac{1}{a} = \dfrac{1}{x}$

19. **a.** Evaluating: $f(e) = 4e^3 \ln e = 4e^3 \approx 80$

b. Finding the derivative: $f'(x) = 4x^3 \cdot \dfrac{1}{x} + 12x^2 \ln x = 4x^2 + 12x^2 \ln x$

Evaluating the derivative: $f'(e) = 4e^2 + 12e^2 \ln e = 16e^2 \approx 118$

c. Setting the derivative equal to 0:

$$4x^2 + 12x^2 \ln x = 0$$
$$4x^2(1 + 3\ln x) = 0$$

$$x = 0 \qquad \text{or} \qquad \ln x = -\dfrac{1}{3}$$
$$x = e^{-1/3}$$

Since $x \neq 0$ for the domain of the logarithm, $x = e^{-1/3}$.

d. Using the point-slope formula:

$$y - 80 = 118(x - e)$$
$$y - 80 = 118x - 321$$
$$y = 118x - 241$$

21. Using implicit differentiation:

$$\dfrac{d}{dx}\left(3x^3 + \ln\left(3xy^2\right)\right) = \dfrac{d}{dx}(2)$$

$$9x^2 + \dfrac{1}{3xy^2}\left(6xyy' + 3y^2\right) = 0$$

$$\dfrac{1}{3xy^2}\left(6xyy' + 3y^2\right) = -9x^2$$

$$6xyy' + 3y^2 = -27x^3y^2$$

$$6xyy' = -27x^3y^2 - 3y^2$$

$$y' = \dfrac{-27x^3y^2 - 3y^2}{6xy} = -\dfrac{9x^3y + y}{2x}$$

23. Using implicit differentiation:

$$\dfrac{d}{dx}\left(2xy^2 + \ln(xy)\right) = \dfrac{d}{dx}(1)$$

$$4xyy' + 2y^2 + \dfrac{1}{xy}(xy' + y) = 0$$

$$4x^2y^2y' + 2xy^3 + xy' + y = 0$$

$$\left(4x^2y^2 + x\right)y' = -2xy^3 - y$$

$$y' = \dfrac{-2xy^3 - y}{4x^2y^2 + x} = -\dfrac{2xy^3 + y}{4x^2y^2 + x}$$

25. Finding the derivative: $f'(x) = \dfrac{1}{x^2 + 2x} \cdot (2x + 2) = \dfrac{2(x+1)}{x(x+2)}$

Evaluating: $f'(3) = \dfrac{2(3+1)}{3(3+2)} = \dfrac{8}{15}$

The function will increase by approximately $\dfrac{8}{15}$ if x increases from 3 to 4.

27. Finding the derivative: $f'(x) = 6x - \dfrac{4}{x^2} \cdot 2x = 6x - \dfrac{8}{x}$

Evaluating: $f'(1) = 6(1) - \dfrac{8}{1} = -2$

The function will decrease by approximately 2 if x increases from 1 to 2.

29. Finding the derivative: $f'(x) = \dfrac{\ln(3x) \cdot \dfrac{1}{4x} \cdot 4 - \ln(4x) \cdot \dfrac{1}{3x} \cdot 3}{\left[\ln(3x)\right]^2} = \dfrac{\ln(3x) - \ln(4x)}{x\left[\ln(3x)\right]^2}$

Evaluating: $f'(1) = \dfrac{\ln(3) - \ln(4)}{\left[\ln(3)\right]^2} \approx -0.238$

The function will decrease by approximately 0.238 if x increases from 1 to 2.

31. The derivative is incorrect (forgot the chain rule).
33. The derivative is incorrect (applied the chain rule incorrectly).
35. The derivative is correct.
37. The derivative is incorrect (applied the chain rule incorrectly).
39. The derivative is incorrect (applied the chain rule incorrectly).

41. Using the power rule and chain rule, the derivative is $5\left[\ln\left(6x^2\right)\right]^4$ multiplied by the derivative of $\ln\left(6x^2\right)$:

$$f'(x) = 5\left[\ln\left(6x^2\right)\right]^4 \cdot \dfrac{1}{6x^2} \cdot 12x$$

43. Finding the derivative: $C'(x) = \dfrac{1{,}000}{3} \cdot \dfrac{1}{3x^2 - 700} \cdot 6x = \dfrac{2{,}000x}{3x^2 - 700}$

Evaluating: $C'(60) = \dfrac{2{,}000(60)}{3(60)^2 - 700} \approx \$11.88$

The manufacturer can expect the cost to increase $11.88 if the production is increased from 60 to 61 units.

45. Finding the derivative: $\dfrac{dR}{dr} = \dfrac{a}{r} - b$

47. Finding the first derivative: $T'(V) = \dfrac{1.44}{V/A_0} \cdot \dfrac{1}{A_0} = \dfrac{1.44}{V}$

Finding the second derivative: $T''(V) = -1.44V^{-2} = -\dfrac{1.44}{V^2}$

Because $V > 0$, the first derivative is positive so the time will increase, but the second derivative is negative so the rate of increase is decreasing. The gallery owner's claim is valid.

49. Finding the derivative: $f'(x) = 5x^2 \cdot \dfrac{1}{x^2} \cdot 2x + 10x\ln\left(x^2\right) = 10x + 10x\ln\left(x^2\right) = 10x\left[1 + \ln\left(x^2\right)\right]$

51. Finding the first derivative: $f'(x) = x \cdot \dfrac{1}{x} + \ln x = 1 + \ln x$

Finding the second derivative: $f''(x) = \dfrac{1}{x}$

53. Using implicit differentiation:

$$\frac{d}{dx}\left(x^2y^3 - \ln\left(x^3y^3\right)\right) = \frac{d}{dx}(10)$$

$$3x^2y^2y' + 2xy^3 - \frac{1}{x^3y^3}\left(3x^3y^2y' + 3x^2y^3\right) = 0$$

$$3x^2y^2y' + 2xy^3 - \frac{1}{xy}\left(3xy' + 3y\right) = 0$$

$$3x^3y^3y' + 2x^2y^4 - 3xy' - 3y = 0$$

$$\left(3x^3y^3 - 3x\right)y' = 3y - 2x^2y^4$$

$$y' = \frac{3y - 2x^2y^4}{3x^3y^3 - 3x} = \frac{y\left(3 - 2x^2y^3\right)}{3x\left(x^2y^3 - 1\right)}$$

55. Simplifying: $e^1 = e$

57. Evaluating: $e^{-0.27} \approx 0.7634$

59. Evaluating: $e^{-0.0578(5)} \approx 0.7490$

61. Differentiating: $\dfrac{dy}{dx} = 2 \cdot 2x = 4x$

63. Differentiating: $\dfrac{dy}{dx} = 3 \cdot 2x + 5 = 6x + 5$

65. Using the point-slope formula:

$$y - 0 = 7(x - 0)$$

$$y = 7x$$

4.4 Differentiating the Natural Exponential Function

1. Differentiating: $f'(x) = e^{3x+6} \cdot 3 = 3e^{3x+6}$

3. Differentiating: $f'(x) = e^{6x} \cdot 6 = 6e^{6x}$

5. Differentiating: $f'(x) = e^{5x^2+4} \cdot 10x = 10xe^{5x^2+4}$

7. Differentiating: $f'(x) = 9e^{3x+1} \cdot 3 = 27e^{3x+1}$

9. Differentiating: $f'(x) = 15x^2 + 6 - 2e^{-3x} \cdot (-3) = 15x^2 + 6 + 6e^{-3x} = 3\left(5x^2 + 2 + 2e^{-3x}\right)$

11. Differentiating: $f'(x) = 5x^2e^{x^2+x} \cdot (2x + 1) + 10xe^{x^2+x} = 5xe^{x^2+x}\left(2x^2 + x + 2\right)$

13. Differentiating: $f'(x) = 3e^{\sqrt[4]{x^2-3x}} \cdot \dfrac{1}{4}\left(x^2 - 3x\right)^{-3/4} \cdot (2x - 3) = \dfrac{3(2x-3)e^{\sqrt[4]{x^2-3x}}}{4\left(x^2 - 3x\right)^{3/4}}$

15. Differentiating: $f'(x) = 2e^{-x} \cdot \dfrac{1}{x^2+2x} \cdot (2x + 2) - 2e^{-x}\ln\left(x^2 + 2x\right) = 2e^{-x}\left[\dfrac{2(x+1)}{x(x+2)} - \ln\left(x^2 + 2x\right)\right]$

17. Differentiating: $f'(x) = 50e^{-0.05x} \cdot (-0.05) = -2.5e^{-0.05x}$

19. Differentiating:

$$f'(x) = \frac{\left(e^x + e^{2x}\right) \cdot 3e^{3x} - e^{3x} \cdot \left(e^x + 2e^{2x}\right)}{\left(e^x + e^{2x}\right)^2}$$

$$= \frac{e^{3x}\left(3e^x + 3e^{2x} - e^x - 2e^{2x}\right)}{\left[e^x\left(1 + e^x\right)\right]^2}$$

$$= \frac{e^{4x}\left(2 + e^x\right)}{e^{2x}\left(1 + e^x\right)^2}$$

$$= \frac{e^{2x}\left(2 + e^x\right)}{\left(1 + e^x\right)^2}$$

21. Using implicit differentiation:

$$\frac{d}{dx}\left(2x^3 + e^{3xy}\right) = \frac{d}{dx}(120)$$

$$6x^2 + e^{3xy}\left(3xy' + 3y\right) = 0$$

$$6x^2 + 3xe^{3xy}y' + 3ye^{3xy} = 0$$

$$3xe^{3xy}y' = -6x^2 - 3ye^{3xy}$$

$$y' = -\frac{6x^2 + 3ye^{3xy}}{3xe^{3xy}} = -\frac{2x^2 + ye^{3xy}}{xe^{3xy}}$$

23. Using implicit differentiation:

$$\frac{d}{dx}\left(e^{xy} + \ln(xy)\right) = \frac{d}{dx}(0)$$

$$e^{xy}\left(xy' + y\right) + \frac{1}{xy}\left(xy' + y\right) = 0$$

$$e^{xy}\left(x^2yy' + xy^2\right) + \left(xy' + y\right) = 0$$

$$x^2ye^{xy}y' + xy^2e^{xy} + xy' + y = 0$$

$$\left(x^2ye^{xy} + x\right)y' = -y - xy^2e^{xy}$$

$$y' = -\frac{y + xy^2e^{xy}}{x^2ye^{xy} + x} = -\frac{y\left(1 + xye^{xy}\right)}{x\left(xye^{xy} + 1\right)} = -\frac{y}{x}$$

25. Finding the first derivative: $f'(x) = e^{x^2+4} \cdot 2x = 2xe^{x^2+4}$

Finding the second derivative: $f''(x) = 2xe^{x^2+4} \cdot 2x + 2e^{x^2+4} = 2e^{x^2+4}\left(2x^2 + 1\right)$

27. Finding the derivative: $f'(x) = 4e^{x^2} \cdot 2x = 8xe^{x^2}$

Evaluating: $f'(1) = 8(1)e^{1^2} = 8e \approx 21.75$

The function will increase by approximately 21.75 if x increases from 1 to 2.

29. Finding the derivative: $f'(x) = 25{,}000e^{-0.065x} \cdot (-0.065) = -1{,}625e^{-0.065x}$

Evaluating: $f'(40) = -1{,}625e^{-0.065(40)} \approx -120.69$

The function will decrease by approximately 120.69 if x increases from 40 to 41.

31.   **a.**   Evaluating: $f(0) = 4(0)e^0 = 0$

      b.   Differentiating: $f'(x) = 4xe^x + 4e^x = 4e^x(x+1)$

            Evaluating: $f'(0) = 4e^0(0+1) = 4$

      c.   Setting the derivative equal to 0:

$$4e^x(x+1) = 0$$

$$x + 1 = 0$$

$$x = -1$$

      d.   Using the point-slope formula:

$$y - 0 = 4(x - 0)$$

$$y = 4x$$

33. The derivative is incorrect (applied the chain rule incorrectly).

35. The derivative is incorrect (applied the chain rule incorrectly).

37. The derivative is correct.

39. The derivative is correct.

41. The derivative is incorrect (applied the exponential rule incorrectly).

43. Finding the derivative: $V'(t) = 65,000e^{-0.46t} \cdot (-0.46) = -29,900e^{-0.46t}$

   a. Evaluating: $V'(1) = -29,900e^{-0.46(1)} \approx -18,875$ sales/day

   b. Evaluating: $V'(6) = -29,900e^{-0.46(6)} \approx -1,892$ sales/day

45. Finding the derivative: $N'(t) = -14,500\left(1 + 65e^{-0.4t}\right)^{-2}\left(65e^{-0.4t} \cdot (-0.4)\right) = \dfrac{377,000}{e^{0.4t}\left(1 + 65e^{-0.4t}\right)^2}$

   Evaluating: $N'(4) = \dfrac{377,000}{e^{0.4(4)}\left(1 + 65e^{-0.4(4)}\right)^2} \approx 382$ more people

47. a. Evaluating: $A(5) = 5e^{-0.0144(5)} \approx 4.7$ mg

   b. Solving the equation:
   $$1.25 = 5e^{-0.0144t}$$
   $$e^{-0.0144t} = 0.25$$
   $$-0.0144t = \ln 0.25$$
   $$t = \frac{\ln 0.25}{-0.0144} \approx 96.3 \text{ hours}$$

   c. Finding the derivative: $A'(t) = 5e^{-0.0144t} \cdot (-0.0144) = -0.072e^{-0.0144t}$

   Evaluating: $A'(1) = -0.072e^{-0.0144(1)} \approx -0.07$ mg/hour

49. Finding the derivative: $A'(t) = 120e^{-0.00012t} \cdot (-0.00012) = -0.0144e^{-0.00012t}$

   Evaluating: $A'(10) = -0.0144e^{-0.00012(10)} \approx -0.0144$ grams

51. a. Evaluating: $P(2) = 100e^{-0.163(2)} \approx 72.2\%$

   b. Finding the derivative: $P'(f) = 100e^{-0.163f} \cdot (-0.163) = -16.3e^{-0.163f}$

   c. Evaluating: $P'(2) = -16.3e^{-0.163(2)} \approx -11.8$ %/foot

   The percentage of surfactants decreases by approximately 11.8% when the feet of clay increases from 2 feet to 3 feet.

53. a. Both models place a $600,000 value on the company's online advertising revenue now.

   b. Azar estimates the decay rate at 1.2%, while Hielo estimates the decay rate at 1.8%.

   c. Finding the derivative:
   $$\frac{d}{dt}\left[R_A(t) - R_H(t)\right] = \frac{d}{dt}\left[600,000e^{-0.012t} - 600,000e^{-0.018t}\right] = 600,000\left(-0.012e^{-0.012t} + 0.018e^{-0.018t}\right)$$

   Evaluating: $\dfrac{d}{dt}\left[R_A(5) - R_H(5)\right] = 600,000\left(-0.012e^{-0.012(5)} + 0.018e^{-0.018(5)}\right) \approx \$3,089.75$

   After 5 years, Azar estimates the online advertising revenue change is $3,089.75 more than Hielo's estimate.

   d. Finding the derivative: $\dfrac{R_H'(t)}{R_A'(t)} = \dfrac{600,000\left(-0.018e^{-0.018t}\right)}{600,000\left(-0.012e^{-0.012t}\right)} = \dfrac{3}{2}e^{-0.006t}$

   Evaluating: $\dfrac{R_H'(5)}{R_A'(5)} = \dfrac{3}{2}e^{-0.006(5)} \approx 1.46$

   After 5 years, Hielo's estimate on the company's online advertising revenue decrease is 1.46 times as much as Azar's.

55. a. In 5 months, the profit prediction of Analyst A is 1.02 times that of Analyst B.

   b. In 5 months, the rate of increase of profit prediction of Analyst A is 1.11 times that of Analyst B.

   c. In 5 months, the ratio of Analyst A's prediction to Analyst B's prediction is increasing by 0.004 units. So the ratio of 1.02 at 5 months would increase to 1.024 in month 6.

57. **a.** Finding the quotient: $\dfrac{R_A(t)}{R_B(t)} = \dfrac{10e^{0.020t}}{10e^{0.017t}} = e^{0.003t}$

Evaluating: $\dfrac{R_A(8)}{R_B(8)} = e^{0.003(8)} \approx 1.02$

In 8 weeks, the revenue prediction of Analyst A will be about 1.02 times that of Analyst B.
The two predictions are very close.

b. Finding the quotient: $\dfrac{R_A'(t)}{R_B'(t)} = \dfrac{10\left(0.020e^{0.020t}\right)}{10\left(0.017e^{0.017t}\right)} = \dfrac{20}{17}e^{0.003t}$

Evaluating: $\dfrac{R_A'(8)}{R_B'(8)} = \dfrac{20}{17}e^{0.003(8)} \approx 1.21$

In 8 weeks, the revenue prediction of Analyst A will be increasing about 1.21 times that of Analyst B's prediction.

c. Finding the derivative: $\left[\dfrac{R_A(t)}{R_B(t)}\right]' = 0.003e^{0.003t}$

Evaluating: $\left.\left[\dfrac{R_A(t)}{R_B(t)}\right]'\right|_{t=8} = 0.003e^{0.003(8)} \approx 0.003$

In 8 weeks, the ratio of A's prediction to B's prediction will increase by 0.003. So the ratio of 1.02 in 8 weeks will increase to approximately 1.023 on week 9.

59. Finding the derivative: $A'(t) = 40.5e^{-0.06931t}(-0.06931) = -2.807055e^{-0.06931t}$

Evaluating: $A'(12) = -2.807055e^{-0.06931(12)} \approx -1.22$ mg/hour

61. Evaluating: $f(0) = 6e^{-0.5(0)+1} + 50 \approx 66.31$

Finding the derivative: $f'(x) = 6e^{-0.5x+1}(-0.5) = -3e^{-0.5x+1}$

Evaluating: $f'(0) = -3e^{-0.5(0)+1} \approx -8.15$

Finding the second derivative: $f''(x) = -3e^{-0.5x+1}(-0.5) = 1.5e^{-0.5x+1}$

Evaluating: $f''(0) = 1.5e^{-0.5(0)+1} \approx 4.08$

63. Finding the derivative: $f'(x) = e^{-0.05x}(1) - 0.05e^{-0.05x}(x+50) = -0.05e^{-0.05x}(x+50-20) = -0.05e^{-0.05x}(x+30)$

Evaluating the derivatives:

$f'(-30.1) = -0.05e^{-0.05(-30.1)}(-30.1+30) \approx 0.0225$

$f'(-30) = -0.05e^{-0.05(-30)}(-30+30) = 0$

$f'(-29.9) = -0.05e^{-0.05(-29.9)}(-29.9+30) \approx -0.0223$

a. The tangent line is horizontal at $x = -30$.

b. Since the slope is positive at $x = -30.1$ and negative at $x = -29.9$, the function appears to reach a maximum at $x = -30$.

65. Writing as a negative exponent: $\dfrac{1}{x^4} = x^{-4}$

67. Rewriting with positive exponents: $\dfrac{x^{-3}}{-3} = -\dfrac{1}{3x^3}$

69. Evaluating and solving for C:

$4 = 1^3 - 2(1)^2 + 2(1) + C$

$4 = 1 + C$

$C = 3$

Chapter 4 Test

1.   **a.**   This function is algebraic.   **b.**   This function is exponential.

2.   **a.**   Expanding the logarithm: $\ln\left(\dfrac{x^3 y^4}{z^2}\right) = \ln\left(x^3\right) + \ln\left(y^4\right) - \ln\left(z^2\right) = 3\ln x + 4\ln y - 2\ln z$

     b.   Compressing the logarithm: $5\ln x - 3\ln y - 4\ln z = \ln\left(x^5\right) - \ln\left(y^3\right) - \ln\left(z^4\right) = \ln\left(\dfrac{x^5}{y^3 z^4}\right)$

3.   Converting to logarithms: $4x = \ln 5$ 4.   Converting to logarithms: $3y - 1 = \ln k$

5.   Converting to exponentials: $3x = e^5$ 6.   Converting to exponentials: $x + 6 = e^{-2}$

7.   Solving the equation: 8.   Solving the equation:

$$e^{2x+1} = 19$$
$$2x + 1 = \ln 19$$
$$2x = \ln 19 - 1$$
$$x = \frac{\ln 19 - 1}{2} \approx 0.9722$$

$$2,500 = 1,500 + 50e^{-0.6x}$$
$$50e^{-0.6x} = 1,000$$
$$e^{-0.6x} = 20$$
$$-0.6x = \ln 20$$
$$x = \frac{\ln 20}{-0.6} \approx -4.9929$$

9.   Finding the derivative: $f'(x) = e^{x^2 - 3x} \cdot (2x - 3)$

     Evaluating at $x = 3$: $f'(3) = e^{3^2 - 3(3)} \cdot (2 \cdot 3 - 3) = 3$

     Using the point-slope formula:
$$y - 1 = 3(x - 3)$$
$$y - 1 = 3x - 9$$
$$y = 3x - 8$$

10.   Finding the derivative: $f'(x) = e^{-3x} \cdot (-3) = -3e^{-3x}$

11.   Finding the derivative: $f'(x) = e^{5x+4} \cdot (5) = 5e^{5x+4}$

12.   Finding the derivative: $f'(x) = x^4 \cdot 6e^{6x} + 4x^3 e^{6x} = 2x^3 e^{6x}(3x + 2)$

13.   Finding the derivative: $f'(x) = \dfrac{1}{8x} \cdot 8 = \dfrac{1}{x}$

14.   Finding the derivative: $f'(x) = \dfrac{1}{(3x - 2)^3} \cdot 3(3x - 2)^2 \cdot 3 = \dfrac{9}{3x - 2}$

15.   Finding the derivative: $f'(x) = 4\left(e^{5x-6}\right)^3 \cdot \left(5e^{5x-6}\right) = 20\left(e^{5x-6}\right)^4 = 20e^{20x-24}$

16.   Finding the derivative: $f'(x) = \dfrac{4}{7x + 7} \cdot 7 + 10e^{7x+7} \cdot 7 = \dfrac{4}{x + 1} + 70e^{7x+7}$

17.   Finding the derivative: $f'(x) = \dfrac{x \cdot 3e^{3x} - \left(e^{3x} + 1\right)}{x^2} = \dfrac{3xe^{3x} - e^{3x} - 1}{x^2}$

18.   Using implicit differentiation:
$$\frac{d}{dx}\left(4x^3 + e^y\right) = \frac{d}{dx}(1)$$
$$12x^2 + e^y(y') = 0$$
$$e^y y' = -12x^2$$
$$y' = -\frac{12x^2}{e^y}$$

19. Finding the first derivative: $f'(x) = e^{3x} \cdot 3 = 3e^{3x}$

   Finding the second derivative: $f''(x) = 3e^{3x} \cdot 3 = 9e^{3x}$

20. Finding the first derivative: $f'(x) = x \cdot \dfrac{1}{x} + \ln x = 1 + \ln x$

   Finding the second derivative: $f''(x) = \dfrac{1}{x}$

21. Finding the derivative: $f'(x) = -3e^{-x}$

   Evaluating: $f'(1) = -3e^{-1} \approx -1.10$

   The function decreases by 1.1 as x changes from 1 to 2.

22. Finding the derivative: $f'(x) = \dfrac{1}{2x - 21} \cdot 2 = \dfrac{2}{2x - 21}$

   Evaluating: $f'(11) = \dfrac{2}{2(11) - 21} = 2$

   The function increases by 2 as x changes from 11 to 12.

23. Solving the equation:
$$25 = 40 - 40e^{-0.07t}$$
$$-40e^{-0.07t} = -15$$
$$e^{-0.07t} = 0.375$$
$$-0.07t = \ln 0.375$$
$$t = \frac{\ln 0.375}{-0.07} \approx 14 \text{ days}$$

24. Solving the equation:
$$175 = 500 - 150e^{0.20t}$$
$$-150e^{0.20t} = -325$$
$$e^{0.20t} = \frac{13}{6}$$
$$0.20t = \ln \frac{13}{6}$$
$$t = \frac{\ln \dfrac{13}{6}}{0.20} \approx 4 \text{ days}$$

25. Finding the derivative: $P'(t) = 220e^{0.08t} \cdot 0.08 = 17.6e^{0.08t}$

   Evaluating: $P'(4) = 17.6e^{0.08(4)} \approx 24.237$ thousand

   The population will be increasing at a rate of 24,237 people/year in 4 years.

26. Finding the derivative: $A'(t) = 10e^{-0.00002876t} \cdot (-0.00002876) = -0.0002876e^{-0.00002876t}$

   Evaluating: $A'(1,000) = -0.0002876e^{-0.00002876(1000)} \approx -0.00028$

   The amount will by decreasing by 0.00028 g/year in 1,000 years.

Chapter 5
Integration: The Language of Accumulation

5.1 Antidifferentiation and the Indefinite Integral

1. Finding the indefinite integral: $\int 6x^5\,dx = 6 \cdot \frac{1}{6}x^6 + C = x^6 + C$

3. Finding the indefinite integral: $\int \frac{1}{x^4}\,dx = \int x^{-4}\,dx = -\frac{1}{3}x^{-3} + C = -\frac{1}{3x^3} + C$

5. Finding the indefinite integral: $\int \frac{8}{\sqrt[5]{x^2}}\,dx = \int 8x^{-2/5}\,dx = 8 \cdot \frac{5}{3}x^{3/5} + C = \frac{40}{3}x^{3/5} + C$

7. Finding the indefinite integral: $\int 6e^x\,dx = 6e^x + C$

9. Finding the indefinite integral: $\int \left(2x^3 + \frac{3}{x^2} + 1\right)dx = \int\left(2x^3 + 3x^{-2} + 1\right)dx = 2 \cdot \frac{1}{4}x^4 - 3x^{-1} + x + C = \frac{1}{2}x^4 - \frac{3}{x} + x + C$

11. Finding the indefinite integral: $\int\left(x^{-3} + x^{-2} - x^{-1}\right)dx = -\frac{1}{2}x^{-2} - x^{-1} - \ln|x| + C = -\frac{1}{2x^2} - \frac{1}{x} - \ln|x| + C$

13. Finding the indefinite integral: $\int\left(0.03 + 0.12x^{-1/2}\right)dx = 0.03x + 0.12 \cdot 2x^{1/2} + C = 0.03x + 0.24x^{1/2} + C$

15. Finding the indefinite integral: $\int\left(5x^4 - 4x^3 + 8x\right)dx = 5 \cdot \frac{1}{5}x^5 - 4 \cdot \frac{1}{4}x^4 + 8 \cdot \frac{1}{2}x^2 + C = x^5 - x^4 + 4x^2 + C$

    Since $f(1) = 9$, we can find the constant: $f(1) = (1)^5 - (1)^4 + 4(1)^2 + C = 4 + C$
    Therefore:
    $$4 + C = 9$$
    $$C = 5$$
    The indefinite integral is $x^5 - x^4 + 4x^2 + 5$.

17. Finding the indefinite integral: $\int\left(3e^x + \frac{4}{x}\right)dx = 3e^x + 4\ln|x| + C$

    Since $f(1) = 5e$, we can find the constant: $f(1) = 3e + 4\ln 1 + C = 3e + C$
    Therefore:
    $$3e + C = 5e$$
    $$C = 2e$$
    The indefinite integral is $3e^x + 4\ln|x| + 2e$.

19. The number of pants the company produces and sells increases by 30 between days 50 and 51. The number of pants sold on day 50 is 550.

21. The number of units that fail increases by 0.008 between 170 and 171 units per minute. The number of units expected to fail is 3 when the rate of production is 170 units per minute.

23. The candidate's favorability rating will decrease by about 0.5% if the negative ads are increased from 10 to 11. The candidate's favorability rating is 45% when the number of negative ads are 10 per week.

25. The number of cars entering the boulevard will increase by 3 from 7:00 AM to 7:01 AM. At 7:00 AM, the number of cars entering the boulevard is 12.

27. Differentiating: $F'(x) = \dfrac{3}{5} \cdot 5x^4 - \dfrac{5}{4} \cdot 4x^3 + 2 \cdot 2x = 3x^4 - 5x^3 + 4x = f(x)$

29. Differentiating:

$$
\begin{aligned}
F'(x) &= \left(x^2+4\right)^3 \cdot 4\left(x^2-5\right)^3(2x) + \left(x^2-5\right)^4 \cdot 3\left(x^2+4\right)^2(2x) \\
&= 8x\left(x^2+4\right)^3\left(x^2-5\right)^3 + 6x\left(x^2-5\right)^4\left(x^2+4\right)^2 \\
&= 2x\left(x^2+4\right)^2\left(x^2-5\right)^3\left[4\left(x^2+4\right)+3\left(x^2-5\right)\right] \\
&= 2x\left(x^2+4\right)^2\left(x^2-5\right)^3\left(4x^2+16+3x^2-15\right) \\
&= 2x\left(x^2+4\right)^2\left(x^2-5\right)^3\left(7x^2+1\right) \\
&= f(x)
\end{aligned}
$$

31. The integral is evaluated correctly.

33. The integral is not evaluated correctly: $\displaystyle\int\left(16x^3+9x^2\right)dx = 16 \cdot \dfrac{1}{4}x^4 + 9 \cdot \dfrac{1}{3}x^3 + C = 4x^4 + 3x^3 + C$

35. The integral is not evaluated correctly: $\displaystyle\int\left(4e^{4x}+x^{-1}\right)dx = 4 \cdot \dfrac{1}{4}e^{4x} + \ln|x| + C = e^{4x} + \ln|x| + C$

37. The integral is not evaluated correctly: $\displaystyle\int\left(e^{2x}+\dfrac{3}{x}+3\right)dx = \dfrac{1}{2}e^{2x} + 3\ln|x| + 3x + C = \dfrac{1}{2}e^{2x} + \ln\left|x^3\right| + 3x + C$

39. The integral is evaluated correctly.

41. **a.** Finding the integral: $\displaystyle\int\left(240t^{1/5}+170\right)dt = 240 \cdot \dfrac{5}{6}t^{6/5} + 170t + C = 200t^{6/5} + 170t + C$

Since $P(0) = 8{,}400$, we can evaluate the constant:

$$P(0) = 200(0)^{6/5} + 170(0) + C$$
$$8{,}400 = C$$

So $P(t) = 200t^{6/5} + 170t + 8{,}400$.

b. Evaluating the function: $P(1) = 200(1)^{6/5} + 170(1) + 8{,}400 = 8{,}770$

c. Evaluating both functions:

$$P'(6) = 240(6)^{1/5} + 170 \approx 513$$

$$P(6) = 200(6)^{6/5} + 170(6) + 8{,}400 \approx 11{,}137$$

The population in 2019 will be 11,137 people, and will grow by 513 people that year.

43. **a.** Evaluating: $M'(10) = -0.018(10)^2 + 0.01(10) = -1.7$

In 10 days, the tumor is decreasing in mass at 1.7 grams per day.

b. Evaluating: $M'(15) = -0.018(15)^2 + 0.01(15) = -3.9$

In 15 days, the tumor is decreasing in mass at 3.9 grams per day.

c. Finding the integral: $\displaystyle\int\left(-0.018t^2+0.01t\right)dt = -0.018 \cdot \dfrac{1}{3}t^3 + 0.01 \cdot \dfrac{1}{2}t^2 + C = -0.006t^3 + 0.005t^2 + C$

Since $M(0) = 200$, we can evaluate the constant:

$$M(0) = -0.006(0)^3 + 0.005(0)^2 + C$$
$$200 = C$$

So $M(t) = -0.006t^3 + 0.005t^2 + 200$.

Evaluating the function: $M(10) = -0.006(10)^3 + 0.005(10)^2 + 200 = 194.5$

In 10 days, the mass of the tumor is 194.5 grams.

d. Evaluating: $M(15) = -0.006(15)^3 + 0.005(15)^2 + 200 = 180.875$

In 15 days, the mass of the tumor is 180.875 grams.

45. **a.** Finding the integral: $\displaystyle\int \frac{0.073}{\sqrt[3]{m}}\,dm = \int 0.073 m^{-1/3}\,dm = 0.073 \cdot \frac{3}{2} m^{2/3} + C = 0.1095 m^{2/3} + C$

Since $A(0) = 0$, we can evaluate the constant:

$$A(0) = 0.1095(0)^{2/3} + C$$
$$0 = C$$

So $A(m) = 0.1095 m^{2/3}$.

Evaluating the function: $A(64) = 0.1095(64)^{2/3} = 1.752$

The surface area of a 64 kg person is 1.752 square meters.

b. Evaluating: $A'(64) = \dfrac{0.073}{\sqrt[3]{64}} = 0.01825$

c. When the person's mass is 64 kg, the surface area is changing at a rate of 0.01825 square meters per kg.

47. Finding the integral: $\displaystyle\int \left(20 - \frac{18,000}{x^2}\right)dx = \int \left(20 - 18,000 x^{-2}\right)dx = 20x + \frac{18,000}{x} + C$

Since $C(50) = 1,360$, we can evaluate the constant:

$$C(50) = 20(50) + \frac{18,000}{50} + C$$
$$1,360 = 1,360 + C$$
$$C = 0$$

So $C(x) = 20x + \dfrac{18,000}{x}$. Evaluating the two functions:

$$C'(30) = 20 - \frac{18,000}{30^2} = 0$$
$$C(30) = 20(30) + \frac{18,000}{30} = 1,200$$

When the chain orders in lots of size 30, its inventory cost is $1,200. Increasing its lot size to 31 will have no effect on its inventory cost.

49. Finding the integral:

$$\int \left(-17.4e^{-1.2t} + 0.28t\right)dt = -\frac{17.4}{-1.2}e^{-1.2t} + 0.28 \cdot \frac{1}{2}t^2 + C = 14.5e^{-1.2t} + 0.14t^2 + C$$

Since $C(0) = 14.5$, we can evaluate the constant:

$$C(0) = 14.5e^{-1.2(0)} + 0.14(0)^2 + C$$
$$14.5 = 14.5 + C$$
$$C = 0$$

So $C(t) = 14.5e^{-1.2t} + 0.14t^2$. Evaluating the two functions:

$$C'(2) = -17.4e^{-1.2(2)} + 0.28(2) \approx -1.018$$
$$C(2) = 14.5e^{-1.2(2)} + 0.14(2)^2 \approx 1.875$$

Two years after the study was completed, the consumption was 1.875 million gallons and decreasing by 1.018 million gallons per year.

51. Finding the integral:

$$\int \left((x-2)^3 + (3x+3)^4\right)dx = \frac{1}{4}(x-2)^4 + \frac{1}{3}\cdot\frac{1}{5}(3x+3)^5 + C = \frac{1}{4}(x-2)^4 + \frac{1}{15}(3x+3)^5 + C$$

The 3rd-degree term is $160x^3$, so the coefficient is 160.

53. The Wolfram Alpha command is: d/dx (int 5x^3 + 2x)

55. Finding the derivative: $\dfrac{du}{dx} = 12x + 4$

57. Solving for du:

$$\dfrac{du}{dx} = 12x + 4$$

$$du = (12x + 4)\,dx$$

59. Finding the derivative: $\dfrac{d}{dx}\left[\dfrac{1}{6}\ln\left(3x^2 + 5\right) + C\right] = \dfrac{1}{6}\cdot\dfrac{6x}{3x^2 + 5} = \dfrac{x}{3x^2 + 5}$

61. Finding the derivative:

$$\dfrac{du}{dx} = 10x$$

$$du = 10x\,dx$$

$$dx = \dfrac{du}{10x}$$

63. Finding the integral: $\dfrac{1}{6}\displaystyle\int\dfrac{1}{u}\,du = \dfrac{1}{6}\ln|u| + C$

65. Finding the derivative: $\dfrac{du}{dx} = \dfrac{2}{5}\cdot\dfrac{5}{2}(x+3)^{3/2} - 2\cdot\dfrac{3}{2}(x+3)^{1/2} = (x+3)^{3/2} - 3(x+3)^{1/2} = (x+3)^{1/2}(x+3-3) = x\sqrt{x+3}$

5.2 Integration by Substitution

1. Let $u = x + 2$, so $\dfrac{du}{dx} = 1$ and thus $du = dx$. Finding the integral:

$$\int(x+2)^4\,dx = \int u^4\,du = \dfrac{1}{5}u^5 + C = \dfrac{1}{5}(x+2)^5 + C$$

3. Let $u = 5x + 1$, so $\dfrac{du}{dx} = 5$ and thus $\dfrac{1}{5}du = dx$. Finding the integral:

$$\int(5x+1)^4\,dx = \dfrac{1}{5}\int u^4\,du = \dfrac{1}{5}\cdot\dfrac{1}{5}u^5 + C = \dfrac{1}{25}(5x+1)^5 + C$$

5. Let $u = x^3 + 2$, so $\dfrac{du}{dx} = 3x^2$ and thus $du = 3x^2\,dx$. Finding the integral:

$$\int 3x^2\left(x^3+2\right)^5\,dx = \int u^5\,du = \dfrac{1}{6}u^6 + C = \dfrac{1}{6}\left(x^3+2\right)^6 + C$$

7. Let $u = x^2 + 4x$, so $\dfrac{du}{dx} = 2x + 4$ and thus $du = (2x+4)\,dx$. Finding the integral:

$$\int(2x+4)\sqrt{x^2+4x}\,dx = \int u^{1/2}\,du = \dfrac{2}{3}u^{3/2} + C = \dfrac{2}{3}\left(x^2+4x\right)^{3/2} + C$$

9. Let $u = 5x^3 - 8x$, so $\dfrac{du}{dx} = 15x^2 - 8$, $du = \left(15x^2 - 8\right)dx$ and thus $4\,du = \left(60x^2 - 32\right)dx$. Finding the integral:

$$\int\left(60x^2 - 32\right)\sqrt{5x^3 - 8x}\,dx = 4\int u^{1/2}\,du = 4\cdot\dfrac{2}{3}u^{3/2} + C = \dfrac{8}{3}\left(5x^3 - 8x\right)^{3/2} + C$$

11. Let $u = 2x^2 + 1$, so $\dfrac{du}{dx} = 4x$ and thus $du = 4x\,dx$. Finding the integral:

$$\int\dfrac{4x}{2x^2+1}\,dx = \int\dfrac{1}{u}\,du = \ln|u| + C = \ln\left(2x^2+1\right) + C$$

13. Let $u = e^{4x} + 2$, so $\dfrac{du}{dx} = 4e^{4x}$ and thus $\dfrac{1}{4}du = e^{4x}dx$. Finding the integral:

$$\int \frac{e^{4x}}{e^{4x}+2}dx = \frac{1}{4}\int \frac{1}{u}du = \frac{1}{4}\ln|u| + C = \frac{1}{4}\ln\left(e^{4x}+2\right) + C$$

15. Let $u = 4x^3 + 5x - 6$, so $\dfrac{du}{dx} = 12x^2 + 5$ and thus $du = \left(12x^2 + 5\right)dx$. Finding the integral:

$$\int \frac{12x^2+5}{4x^3+5x-6}dx = \int \frac{1}{u}du = \ln|u| + C = \ln\left|4x^3+5x-6\right| + C$$

17. Let $u = \ln x + 7$, so $\dfrac{du}{dx} = \dfrac{1}{x}$ and thus $du = \dfrac{1}{x}dx$. Finding the integral:

$$\int \frac{\ln x + 7}{x}dx = \int u\,du = \frac{1}{2}u^2 + C = \frac{1}{2}(\ln x + 7)^2 + C$$

19. Let $u = 5x - 4$, so $\dfrac{du}{dx} = 5$ and thus $du = 5dx$. Solving for x:

$$u = 5x - 4$$
$$u + 4 = 5x$$
$$x = \frac{1}{5}(u + 4)$$

Finding the integral:

$$\int \frac{5x}{5x-4}dx = \int \frac{1}{u}\cdot\frac{1}{5}(u+4)du$$
$$= \frac{1}{5}\int \left(1 + \frac{4}{u}\right)du$$
$$= \frac{1}{5}\left(u + 4\ln|u|\right) + C$$
$$= \frac{1}{5}\left(5x - 4 + 4\ln|5x-4|\right) + C$$
$$= x + \frac{4}{5}\ln|5x-4| + C$$

21. Let $u = -0.02x$, so $\dfrac{du}{dx} = -0.02$ and thus $\dfrac{du}{-0.02} = dx$. Finding the integral:

$$\int e^{-0.02x}\,dx = \int \frac{e^u}{-0.02}du = -50e^u + C = -50e^{-0.02x} + C$$

23. Let $u = 0.15x$, so $\dfrac{du}{dx} = 0.15$ and thus $\dfrac{du}{0.15} = dx$. Finding the integral:

$$\int 45e^{0.15x}\,dx = 45\int \frac{e^u}{0.15}du = 300e^u + C = 300e^{0.15x} + C$$

25. To do this integration, we will split the integral into two separate integrals: $\int \left(e^{3x} + 12e^{-3x}\right)dx = \int e^{3x}\,dx + 12\int e^{-3x}\,dx$

For the first integral, let $u = 3x$, so $\dfrac{du}{dx} = 3$ and thus $\dfrac{1}{3}du = dx$: $\int e^{3x}\,dx = \dfrac{1}{3}\int e^u\,du = \dfrac{1}{3}e^u + C = \dfrac{1}{3}e^{3x} + C$

For the second integral, let $u = -3x$, so $\dfrac{du}{dx} = -3$ and thus $-\dfrac{1}{3}du = dx$: $\int e^{-3x}\,dx = -\dfrac{1}{3}\int e^u\,du = -\dfrac{1}{3}e^u + C = -\dfrac{1}{3}e^{-3x} + C$

Therefore, the integration is: $\int \left(e^{3x} + 12e^{-3x}\right)dx = \int e^{3x}\,dx + 12\int e^{-3x}\,dx = \dfrac{1}{3}e^{3x} + 12\left(-\dfrac{1}{3}e^{-3x}\right) + C = \dfrac{1}{3}e^{3x} - 4e^{-3x} + C$

27. Let $u = ax$, so $\dfrac{du}{dx} = a$ and thus $\dfrac{1}{a}du = dx$. Finding the integral:

$$\int e^{ax}\,dx = \frac{1}{a}\int e^u\,du = \frac{1}{a}e^u + C = \frac{1}{a}e^{ax} + C$$

29. Let $u = 5x^2 + 4$, so $\dfrac{du}{dx} = 10x$ and thus $du = 10x\,dx$. Thus $A = 10x$.

31. Let $u = 4x^3 + x^2$, so $\dfrac{du}{dx} = 12x^2 + 2x$ and thus $\dfrac{1}{2}du = \left(6x^2 + x\right)dx$. Thus $A = 6x^2 + x$.

33. Let $u = x^2 + 2x$, so $\dfrac{du}{dx} = 2x + 2$ and thus $5du = \left(10x + 10\right)dx$. Thus $A = 10x + 10$.

35. Let $u = \ln\left(x^2\right)$, so $\dfrac{du}{dx} = \dfrac{1}{x^2}\cdot 2x = \dfrac{2}{x}$ and thus $du = \dfrac{2}{x}dx$. Thus $A = \dfrac{2}{x}$.

37. **a.** Let $u = 2t + 3$, so $\dfrac{du}{dt} = 2$ and thus $\dfrac{1}{2}du = dt$. Finding the integral:

$$\int \frac{3{,}000}{10t + 15}\,dt = \int \frac{600}{2t + 3}\,dt = \frac{600}{2}\int \frac{1}{u}\,du = 300\ln|u| + C = 300\ln(2t + 3) + C$$

Since $N(0) = 0$, we can find C:

$$N(0) = 300\ln(0 + 3) + C$$
$$0 = 300\ln 3 + C$$
$$C = -300\ln 3$$

So $N(t) = 300\ln(2t + 3) - 300\ln 3 = 300\ln\left(\dfrac{2t + 3}{3}\right)$.

b. Evaluating: $N(5) = 300\ln\left(\dfrac{13}{3}\right) \approx 439.9$

Approximately 439,900 antibodies are produced 5 days after vaccination.

39. **a.** Let $u = -0.004t$, so $\dfrac{du}{dt} = -0.004$ and thus $-\dfrac{1}{0.004}du = dt$. Finding the integral:

$$\int -1{,}580e^{-0.004t}\,dt = \frac{-1{,}580}{-0.004}\int e^u\,du = 395{,}000e^u + C = 395{,}000e^{-0.004t} + C$$

Since $N(10) = 370{,}000$, we can find C:

$$N(10) = 395{,}000e^{-0.004(10)} + C$$
$$370 = 379{,}511.83 + C$$
$$C = -9{,}511.83$$

So $N(t) = 395{,}000e^{-0.004t} - 9{,}511.83$.

b. Evaluating: $N(24) = 395{,}000e^{-0.004(24)} - 9{,}511.83 \approx 349{,}331.46$

The resale value is $349,331.46 after 24 months.

c. Finding the limit: $\displaystyle\lim_{t\to\infty}\left(395{,}000e^{-0.004t} - 9{,}511.83\right) = \lim_{t\to\infty}\left(\dfrac{395{,}000}{e^{0.004t}} - 9{,}511.83\right) = 0$

The resale value will approach 0 in the long run (it cannot be negative).

41. **a.** Let $u = -0.32t$, so $\dfrac{du}{dt} = -0.32$ and thus $-\dfrac{1}{0.32}du = dt$. Finding the integral:

$$\int -6{,}875e^{-0.32t}\,dt = \dfrac{-6{,}875}{-0.32}\int e^u\,du = 21{,}484.375e^u + C = 21{,}484.375e^{-0.32t} + C$$

Since $N(0) = 55{,}000$, we can find C:

$$N(0) = 21{,}484.375e^{-0.32(0)} + C$$
$$55{,}000 = 21{,}484.375 + C$$
$$C = 33{,}515.625$$

So $N(t) = 21{,}484.375e^{-0.32t} + 33{,}515.625$.

b. Evaluating: $N(10) = 21{,}484.375e^{-0.32(10)} + 33{,}515.625 \approx 34{,}391.38$

The number of daily sales is \$34,391.38 after 10 days.

c. Finding the limit: $\lim\limits_{t\to\infty}\left(21{,}484.375e^{-0.32t} + 33{,}515.625\right) = \lim\limits_{t\to\infty}\left(\dfrac{21{,}484.375}{e^{0.32t}} + 33{,}515.625\right) \approx 33{,}515.63$

The limiting number of sales in the long run is \$33,515.63.

43. **a.** Let $u = 1 + e^{18t}$, so $\dfrac{du}{dt} = 18e^{18t}$ and thus $\dfrac{1}{18}du = e^{18t}\,dt$. Finding the integral:

$$\int \dfrac{126e^{18t}}{1+e^{18t}}\,dt = \dfrac{126}{18}\int \dfrac{1}{u}\,du = 7\ln|u| + C = 7\ln\left(1 + e^{18t}\right) + C$$

Since $P(0) = 28{,}000$, we can find C:

$$P(0) = 7\ln\left(1 + e^{18(0)}\right) + C$$
$$28{,}000 = 7\ln 2 + C$$
$$C \approx 27{,}995.148$$

So $P(t) = 7\ln\left(1 + e^{18t}\right) + 27{,}995.148$.

b. Evaluating: $P(10) = 7\ln\left(1 + e^{18(10)}\right) + 27{,}995.148 \approx 29{,}255$

The population will be 29,255 people 10 months from now.

c. Evaluating: $P'(5) = \dfrac{126e^{18(5)}}{1+e^{18(5)}} \approx 126$

The population will increase by 126 people from month 5 to month 6.

45. **a.** Let $u = 100 - x$, so $\dfrac{du}{dx} = -1$ and thus $-du = dx$. Solving for x:

$$u = 100 - x$$
$$x = 100 - u$$
$$200 - 3x = 200 - 3(100 - u)$$
$$200 - 3x = 3u - 100$$

Finding the integral:

$$\int \frac{1}{3} \cdot \frac{200-3x}{\sqrt{100-x}}\,dx = -\frac{1}{3}\int \frac{3u-100}{\sqrt{u}}\,du$$

$$= -\frac{1}{3}\int\left(3u^{1/2}-100u^{-1/2}\right)du$$

$$= -\frac{1}{3}\left(2u^{3/2}-200u^{1/2}\right)+C$$

$$= -\frac{2}{3}(100-x)^{3/2}+\frac{200}{3}(100-x)^{1/2}+C$$

$$= \frac{2}{3}(100-x)^{1/2}(x-100+100)+C$$

$$= \frac{2}{3}x\sqrt{100-x}+C$$

Since $R(0)=0$, we can find C:

$$R(0)=\frac{2}{3}(0)\sqrt{100-0}+C$$

$$0=C$$

So $R(x)=\frac{2}{3}x\sqrt{100-x}$.

b. Evaluating: $R(51)=\frac{2}{3}(51)\sqrt{100-51}=238$

The company's revenue is \$238,000 after 51 weeks.

47. a. Let $u=t+3$, so $\frac{du}{dt}=1$ and thus $du=dt$. Finding the integral:

$$\int \frac{1,020}{(t+3)^2}\,dt = 1,020\int u^{-2}\,du = 1,020\left(-u^{-1}\right)+C = -\frac{1,020}{t+3}+C$$

Since $V(0)=0$, we can find C:

$$V(0)=-\frac{1,020}{0+3}+C$$

$$0=-340+C$$

$$C=340$$

So $V(t)=-\frac{1,020}{t+3}+340$.

b. Evaluating: $V(2)=-\frac{1,020}{2+3}+340=136$

The dragster is traveling at 136 mph after 2 seconds into its run.

49. Let $u=x^3-7$, so $\frac{du}{dx}=3x^2$ and thus $du=3x^2\,dx$. Finding the integral:

$$\int 3x^2\left(x^3-7\right)^5\,dx = \int u^5\,du = \frac{1}{6}u^6+C = \frac{1}{6}\left(x^3-7\right)^6+C$$

51. Let $u=4+5\ln x$, so $\frac{du}{dx}=\frac{5}{x}$ and thus $\frac{1}{5}du=\frac{1}{x}\,dx$. Finding the integral:

$$\int \frac{\sqrt{4+5\ln x}}{x}\,dx = \frac{1}{5}\int u^{1/2}\,du = \frac{1}{5}\cdot\frac{2}{3}u^{3/2}+C = \frac{2}{15}(4+5\ln x)^{3/2}+C$$

53. Let $u=ax+b$, so $\frac{du}{dx}=a$ and thus $\frac{1}{a}du=dx$. Finding the integral: $\int e^{ax+b}\,dx=\frac{1}{a}\int e^u\,du=\frac{1}{a}e^u+C=\frac{1}{a}e^{ax+b}+C$

55. Evaluating: $T(7)-T(3)=\left(40e^{1.2(7)}+C\right)-\left(40e^{1.2(3)}+C\right)=40e^{8.4}-40e^{3.6}\approx 176,418.74$

57. Let $u = x - 3$, so $\dfrac{du}{dx} = 1$ and thus $du = dx$. Finding the integral: $\displaystyle\int \frac{1}{x-3}\,dx = \int\frac{1}{u}\,du = \ln|u| + C = \ln|x-3| + C$

59. Finding the integral: $\displaystyle\int\frac{1}{u}\,du = \ln|u| + C$

Subtracting the values: $\left(\ln|7| + C\right) - \left(\ln|1| + C\right) = \ln 7 - \ln 1 = \ln 7 \approx 1.9459$

5.3 The Definite Integral

1. Finding the integral: $\displaystyle\int_0^2 \left(3x^2 + 4x - 1\right)dx = x^3 + 2x^2 - x\Big|_0^2 = (8 + 8 - 2) - (0 + 0 - 0) = 14$

3. Finding the integral: $\displaystyle\int_0^4 \left(\frac{3}{2}x^2 - 10x + 8\right)dx = \frac{1}{2}x^3 - 5x^2 + 8x\Big|_0^4 = (32 - 80 + 32) - (0 - 0 + 0) = -16$

5. Finding the integral: $\displaystyle\int_2^5 \left(4x^2 + 4x + 1\right)dx = \frac{4}{3}x^3 + 2x^2 + x\Big|_2^5 = \left(\frac{500}{3} + 50 + 5\right) - \left(\frac{32}{3} + 8 + 2\right) = 201$

7. Finding the integral: $\displaystyle\int_5^{10} \left(5x^2 + x\right)dx = \frac{5}{3}x^3 + \frac{1}{2}x^2\Big|_5^{10} = \left(\frac{5{,}000}{3} + 50\right) - \left(\frac{625}{3} + \frac{25}{2}\right) = \frac{8{,}975}{6}$

9. Finding the integral: $\displaystyle\int_0^1 \left(4x^3 - 3x^2\right)dx = x^4 - x^3\Big|_0^1 = (1 - 1) - (0 - 0) = 0$

11. Let $u = 2x$, so $\dfrac{du}{dx} = 2$ and thus $\dfrac{1}{2}du = dx$. When $x = 0$, $u = 0$, and when $x = 4$, $u = 8$. Finding the integral:

$$\int_0^4 4e^{2x}\,dx = \frac{4}{2}\int_0^8 e^u\,du = 2e^u\Big|_0^8 = \left(2e^8\right) - \left(2e^0\right) = 2\left(e^8 - 1\right)$$

13. Let $u = -5x$, so $\dfrac{du}{dx} = -5$ and thus $-\dfrac{1}{5}du = dx$. When $x = 0$, $u = 0$, and when $x = 2$, $u = -10$. Finding the integral:

$$\int_0^2 10e^{-5x}\,dx = \frac{10}{-5}\int_0^{-10} e^u\,du = -2e^u\Big|_0^{-10} = \left(-2e^{-10}\right) - \left(-2e^0\right) = 2\left(1 - e^{-10}\right)$$

15. Let $u = 5x$, so $\dfrac{du}{dx} = 5$ and thus $\dfrac{1}{5}du = dx$. When $x = 0$, $u = 0$, and when $x = \ln e$, $u = 5\ln e$. Finding the integral:

$$\int_0^{\ln e} e^{5x}\,dx = \frac{1}{5}\int_0^{5\ln e} e^u\,du = \frac{1}{5}e^u\Big|_0^{\ln e^5} = \left(\frac{1}{5}e^{\ln e^5}\right) - \left(\frac{1}{5}e^0\right) = \frac{1}{5}\left(e^5 - 1\right)$$

17. Let $u = -2x$, so $\dfrac{du}{dx} = -2$ and thus $-\dfrac{1}{2}du = dx$. When $x = 0$, $u = 0$, and when $x = \ln e$, $u = -2\ln e$.

Finding the integral: $\displaystyle\int_0^{\ln e} -8e^{-2x}\,dx = \frac{-8}{-2}\int_0^{-2\ln e} e^u\,du = 4e^u\Big|_0^{\ln e^{-2}} = \left(4e^{\ln e^{-2}}\right) - \left(4e^0\right) = 4\left(e^{-2} - 1\right)$

19. Finding the integral: $\displaystyle\int_1^e \frac{3}{x}\,dx = 3\ln|x|\Big|_1^e = 3\ln e - 3\ln 1 = 3$

21. Finding the integral: $\displaystyle\int_1^2 \frac{3}{x^2}\,dx = \int_1^2 3x^{-2}\,dx = -\frac{3}{x}\Big|_1^2 = -\frac{3}{2} - (-3) = \frac{3}{2}$

23. Finding the integral: $\displaystyle\int_0^5 7\,dx = 7x\Big|_0^5 = 35 - 0 = 35$

25. Let $u = x - 3$, so $\dfrac{du}{dx} = 1$ and thus $du = dx$. When $x = 4$, $u = 1$, and when $x = 5$, $u = 2$. Finding the integral:

$$\int_4^5 \frac{2}{x-3}\,dx = \int_1^2 \frac{2}{u}\,du = 2\ln|u|\Big|_1^2 = 2\ln 2 - 2\ln 1 = 2\ln 2 = \ln 2^2 = \ln 4$$

27. Let $u = x^2 - 3$, so $\dfrac{du}{dx} = 2x$ and thus $du = 2x\,dx$. When $x = 2$, $u = 1$, and when $x = 3$, $u = 6$. Finding the integral:

$$\int_2^3 \frac{2x}{x^2 - 3}\,dx = \int_1^6 \frac{1}{u}\,du = \ln|u|\Big|_1^6 = \ln 6 - \ln 1 = \ln 6$$

29. Let $u = 6x^2 + 2x$, so $\dfrac{du}{dx} = 12x + 2$ and thus $du = (12x + 2)\,dx$. When $x = 1$, $u = 8$, and when $x = 2$, $u = 28$.

Finding the integral: $\displaystyle\int_1^2 \frac{12x + 2}{6x^2 + 2x}\,dx = \int_8^{28} \frac{1}{u}\,du = \ln|u|\Big|_8^{28} = \ln 28 - \ln 8 = \ln\frac{28}{8} = \ln\frac{7}{2}$

31. Let $u = x^2 + 6$, so $\dfrac{du}{dx} = 2x$ and thus $\dfrac{1}{2}du = x\,dx$. When $x = 0$, $u = 6$, and when $x = 5$, $u = 31$.

Finding the integral: $\displaystyle\int_0^5 \frac{6x}{x^2 + 6}\,dx = \frac{6}{2}\int_6^{31} \frac{1}{u}\,du = 3\ln|u|\Big|_6^{31} = 3\ln 31 - 3\ln 6 = 3\ln\frac{31}{6}$

33. Let $u = 3x^2 + x$, so $\dfrac{du}{dx} = 6x + 1$ and thus $du = (6x + 1)\,dx$. When $x = 0$, $u = 0$, and when $x = 1$, $u = 4$.

Finding the integral: $\displaystyle\int_0^1 (6x + 1)e^{3x^2 + x}\,dx = \int_0^4 e^u\,du = e^u\Big|_0^4 = e^4 - e^0 = e^4 - 1$

35. Finding the integral: $\displaystyle\int_{\ln 3}^{\ln 6} 3e^x\,dx = 3e^x\Big|_{\ln 3}^{\ln 6} = 3\left(e^{\ln 6} - e^{\ln 3}\right) = 3(6 - 3) = 9$

37. Let $u = \ln x$, so $\dfrac{du}{dx} = \dfrac{1}{x}$ and thus $du = \dfrac{1}{x}dx$. When $x = 2$, $u = \ln 2$, and when $x = 4$, $u = \ln 4$.

Finding the integral: $\displaystyle\int_2^4 \frac{4}{x\ln^2 x}\,dx = 4\int_{\ln 2}^{\ln 4} u^{-2}\,du = \frac{-4}{u}\Big|_{\ln 2}^{\ln 4} = -4\left(\frac{1}{\ln 4} - \frac{1}{\ln 2}\right) = \frac{-4}{2\ln 2} + \frac{4}{\ln 2} = \frac{-2}{\ln 2} + \frac{4}{\ln 2} = \frac{2}{\ln 2}$

39. The error occurs in step 2, as the terms are reversed. The corrected step is: $\left(\dfrac{3(6)^2}{2} - 4(6)\right) - \left(\dfrac{3(2)^2}{2} - 4(2)\right)$

41. The error occurs in step 1, as the function was differentiated rather than integrated. The corrected step is:

$$\int_2^3 3e^{3x}\,dx = e^{3x}\Big|_2^3 = e^9 - e^6$$

43. This represents the total amount of money spent on advertising from month a to month b.
45. This represents the total gallons of gasoline pumped between day a and day b.
47. This represents the total amount of chemical leaked into the lake between hour a and hour b.
49. This represents the total change in wealth between age a and age b.
51. This represents the total income stream change from year a to year b.
53. This represents the total depreciation from month a to month b.
55. This represents the total change in probability from time a to time b.

57. Finding the integral: $\displaystyle\int_{100}^{120} (100 - 0.6x)\,dx = 100x - 0.3x^2\Big|_{100}^{120} = \left(100 \cdot 120 - 0.3 \cdot 120^2\right) - \left(100 \cdot 100 - 0.3 \cdot 100^2\right) = 680$

It will cost \$680 to increase production from 100 to 120 lamps.

59. **a.** Finding the integral:

$$\int_0^{10} \left(6t^4 - 720t^3 + 21{,}600t^2\right)dt = \frac{6}{5}t^5 - 180t^4 + 7{,}200t^3\Big|_0^{10}$$

$$= \left(\frac{6}{5} \cdot 10^5 - 180 \cdot 10^4 + 7{,}200 \cdot 10^3\right) - (0)$$

$$= 5{,}520{,}000$$

A total of 5,520,000 gallons are extracted during the first 10 days of operation.

b. Finding the integral:

$$\int_0^{60}\left(6t^4-720t^3+21{,}600t^2\right)dt=\frac{6}{5}t^5-180t^4+7{,}200t^3\Big|_0^{60}$$

$$=\left(\frac{6}{5}\cdot60^5-180\cdot60^4+7{,}200\cdot60^3\right)-(0)$$

$$=155{,}520{,}000$$

A total of 155,520,000 gallons are extracted during the first 60 days of operation.

61. Finding the integral:

$$\int_0^5\left(-x^4+11x^3-39x^2+45x\right)dx=-\frac{1}{5}x^5+\frac{11}{4}x^4-13x^3+\frac{45}{2}x^2\Big|_0^5$$

$$=\left(-\frac{1}{5}\cdot5^5+\frac{11}{4}\cdot5^4-13\cdot5^3+\frac{45}{2}\cdot5^2\right)-(0)$$

$$=31.25$$

The total revenue for the five year period was $31.25 million.

63. Let $u=t+3$, so $\dfrac{du}{dt}=1$ and thus $du=dt$. When $t=0,u=3$, and when $t=4,u=7$. Note that $t=u-3$.

Finding the integral:

$$\int_0^4\frac{340t}{t+3}dt=340\int_3^7\frac{u-3}{u}du$$

$$=340\int_3^7\left(1-\frac{3}{u}\right)du$$

$$=340\left(u-3\ln|u|\right)\Big|_3^7$$

$$=340(7-3\ln7)-340(3-3\ln3)$$

$$=340(4-3\ln7+3\ln3)$$

$$\approx495.8$$

Her dragster traveled 495.8 feet in the first 4 seconds.

65. **a.** Finding the derivative: $C'(x)=\dfrac{(x+1)(400)-(400x+1{,}000)(1)}{(x+1)^2}=\dfrac{200(2x+2-2x-5)}{(x+1)^2}=\dfrac{-600}{(x+1)^2}$

For $0\le x\le100$, this derivative is negative, so the cost is decreasing.

b. Finding the limit: $\displaystyle\lim_{x\to\infty}\frac{400x+1{,}000}{x+1}=\lim_{x\to\infty}\frac{400x}{x}=400$

The minimum cost is the limiting value of $400.

c. Let $u=x+1$, so $\dfrac{du}{dx}=1$ and thus $du=dx$. When $x=0,u=1$, and when $x=100,u=101$. Note that $x=u-1$.

Finding the integral:

$$\int_0^{100}\frac{400x+1{,}000}{x+1}dx=\int_1^{101}\frac{400(u-1)+1{,}000}{u}du$$

$$=\int_1^{101}\frac{400u+600}{u}du$$

$$=\int_1^{101}\left(400+\frac{600}{u}\right)du$$

$$=400u+600\ln|u|\Big|_1^{101}$$

$$=(400\cdot101+600\ln101)-400$$

$$\approx\$42{,}769$$

The total cost to the company is approximately $42,769 as it increases production from 0 to 100 instruments.

67. **a.** Finding the first and second derivatives:

$$R'(x) = -480\left(-0.002e^{-0.002x}\right) = 0.96e^{-0.002x}$$

$$R''(x) = 0.96\left(-0.002e^{-0.002x}\right) = -0.00192e^{-0.002x}$$

Since the first derivative is positive and the second derivative is negative, the monthly revenue is increasing at a decreasing rate, supporting the analyst's claim.

b. Let $u = -0.002x$, so $\dfrac{du}{dx} = -0.002$ and thus $-\dfrac{1}{0.002}du = dx$. When $x = 0$, $u = 0$, and when $x = 1{,}000$, $u = -2$.

Finding the integral:

$$\int_0^{1{,}000}\left(800 - 480e^{-0.002x}\right)dx = \int_0^{1{,}000}800\,dx - 480\int_0^{1{,}000}e^{-0.002x}\,dx$$

$$= 800x\Big|_0^{1{,}000} - \frac{480}{-0.002}\int_0^{-2}e^u\,du$$

$$= 800x\Big|_0^{1{,}000} + 240{,}000\,e^u\Big|_0^{-2}$$

$$= 800(1{,}000) - 800(0) + 240{,}000e^{-2} - 240{,}000$$

$$\approx \$592{,}480$$

The total revenue over the year will be approximately \$582,480, which is more than \$500,000, supporting the analyst's claim.

69. **a.** Let $u = 1.5t^2 + 8t$, so $\dfrac{du}{dt} = 3t + 8$ and thus $\dfrac{1}{2}du = (1.5t + 4)dt$. When $t = 0$, $u = 0$, and when $t = 1$, $u = 9.5$.

Finding the integral:

$$\int_0^1(1.5t + 4)\sqrt{1.5t^2 + 8t}\,dt = \frac{1}{2}\int_0^{9.5}u^{1/2}\,dt = \frac{1}{2}\cdot\frac{2}{3}u^{3/2}\Big|_0^{9.5} = \frac{1}{3}(9.5)^{3/2} - 0 \approx 9.760$$

The expected cost to maintain the machine in its first year is \$9,760.

b. Let $u = 1.5t^2 + 8t$, so $\dfrac{du}{dt} = 3t + 8$ and thus $\dfrac{1}{2}du = (1.5t + 4)dt$. When $t = 9$, $u = 193.5$, and when $t = 10$, $u = 230$.

Finding the integral:

$$\int_9^{10}(1.5t + 4)\sqrt{1.5t^2 + 8t}\,dt = \frac{1}{2}\int_{193.5}^{230}u^{1/2}\,dt = \frac{1}{2}\cdot\frac{2}{3}u^{3/2}\Big|_{193.5}^{230} = \frac{1}{3}(230)^{3/2} - \frac{1}{3}(193.5)^{3/2} \approx 265.485$$

The expected cost to maintain the machine in its last year is \$265,485, which exceeds the \$250,000 limit. The company should scrap the machine.

c. Comparing the two costs: $\dfrac{265{,}485}{9{,}760} \approx 27.2$

It is 27.2 times as expensive to maintain the machine in the last year compared to the first year.

71. Let $u = 0.3x^2 + 200$, so $\dfrac{du}{dx} = 0.6$ and thus $\dfrac{1}{0.6}du = x\,dx$. When $x = 0$, $u = 200$, and when $x = 30$, $u = 470$.

Finding the integral: $\displaystyle\int_0^{30}\frac{400x}{0.3x^2 + 200}\,dx = \frac{400}{0.6}\int_{200}^{470}\frac{1}{u}\,du = \frac{2{,}000}{3}\ln|u|\Big|_{200}^{470} = \frac{2{,}000}{3}(\ln 470 - \ln 200) \approx 569.61$

73. Let $u = -0.03t$, so $\dfrac{du}{dt} = -0.03$ and thus $-\dfrac{1}{0.03}du = dt$. When $t = 0, u = 0$, and when $t = 5, u = -0.15$.

Finding the integral:

$$\int_0^5 \left(355 - 145e^{-0.03t}\right)dt = \int_0^5 355\,dt - 145\int_0^5 e^{-0.03t}\,dt$$

$$= 355t\Big|_0^5 - \frac{145}{-0.03}\int_0^{-0.15} e^u\,du$$

$$= 355t\Big|_0^5 + \frac{14{,}500}{3}e^u\Big|_0^{-0.15}$$

$$= 355(5) - 355(0) + \frac{14{,}500}{3}e^{-0.15} - \frac{14{,}500}{3}$$

$$\approx 1{,}102$$

The total number of units an average worker can assemble after 5 weeks of experience is approximately 1,102 units.

Evaluating the function: $N(5) = 355 - 145e^{-0.03(5)} \approx 230$

After 5 weeks of experience, the average worker can assemble 230 units per week.

75. Finding the two index values:

$$\text{A: } 1 - \int_0^1 x^{0.82}\,dx = 1 - \frac{x^{1.82}}{1.82}\bigg|_0^1 = 1 - \frac{1}{1.82} \approx 0.4505$$

$$\text{B: } 1 - \int_0^1 x^{0.67}\,dx = 1 - \frac{x^{1.67}}{1.67}\bigg|_0^1 = 1 - \frac{1}{1.67} \approx 0.4012$$

Since the Gini index for Country B is closer to 0, the money in Country B is more evenly distributed than in Country A.

77. Finding the integral:

$$\int_1^6 \left(0.3x^3 - 1.8x^2 + 15\right)dx = 0.075x^4 - 0.6x^3 + 15x\Big|_1^6$$

$$= \left(0.075 \cdot 6^4 - 0.6 \cdot 6^3 + 15 \cdot 6\right) - \left(0.075 \cdot 1^4 - 0.6 \cdot 1^3 + 15 \cdot 1\right)$$

$$= 43.125$$

79. Evaluating the integrals:

$$\int_2^4 -\left(x^2 - 5x + 4\right)dx = -\frac{1}{3}x^3 + \frac{5}{2}x^2 - 4x\bigg|_2^4 = \left(-\frac{1}{3}\cdot 4^3 + \frac{5}{2}\cdot 4^2 - 4\cdot 4\right) - \left(-\frac{1}{3}\cdot 2^3 + \frac{5}{2}\cdot 2^2 - 4\cdot 2\right) = \frac{10}{3}$$

$$\int_4^5 \left(x^2 - 5x + 4\right)dx = \frac{1}{3}x^3 - \frac{5}{2}x^2 + 4x\bigg|_4^5 = \left(\frac{1}{3}\cdot 5^3 - \frac{5}{2}\cdot 5^2 + 4\cdot 5\right) - \left(\frac{1}{3}\cdot 4^3 - \frac{5}{2}\cdot 4^2 + 4\cdot 4\right) = \frac{11}{6}$$

Therefore: $\int_2^4 -\left(x^2 - 5x + 4\right)dx + \int_4^5 \left(x^2 - 5x + 4\right)dx = \dfrac{10}{3} + \dfrac{11}{6} = \dfrac{31}{6}$

5.4 The Definite Integral and Area

1. Finding the area: $\int_0^3 \left(-6x^2 + 26x\right)dx = -2x^3 + 13x^2\Big|_0^3 = (-54 + 117) - 0 = 63$ units^2

3. Finding the area: $\int_0^2 \left(3x^2 - 3x + 5\right)dx = x^3 - \frac{3}{2}x^2 + 5x\Big|_0^2 = (8 - 6 + 10) - 0 = 12$ units^2

5. Finding the area: $\int_1^4 \frac{1}{x}dx = \ln|x|\Big|_1^4 = \ln 4 - \ln 1 = \ln 4$ units^2

7. Let $u = 0.2x$, so $\frac{du}{dx} = 0.2$ and thus $\frac{1}{0.2}du = dx$. When $x = 0$, $u = 0$, and when $x = 4$, $u = 0.8$.

Finding the area: $\int_0^4 2{,}500e^{0.2x}dx = \frac{2{,}500}{0.2}\int_0^{0.8} e^u du = 12{,}500e^u\Big|_0^{0.8} = 12{,}500\left(e^{0.8} - 1\right) \approx 15{,}319.26$ units^2

9. Finding the area: $\int_1^4 2\sqrt{x}\,dx = \int_1^4 2x^{1/2}dx = 2 \cdot \frac{2}{3}x^{3/2}\Big|_1^4 = \frac{4}{3}(8 - 1) = \frac{28}{3}$ units^2

11. Let $u = x^2 + 8$, so $\frac{du}{dx} = 2x$ and thus $\frac{1}{2}du = xdx$. When $x = 3$, $u = 17$, and when $x = 5$, $u = 33$.

Finding the area: $\int_3^5 \frac{x}{x^2 + 8}dx = \frac{1}{2}\int_{17}^{33} \frac{1}{u}du = \frac{1}{2}\ln|u|\Big|_{17}^{33} = \frac{1}{2}(\ln 33 - \ln 17)^4 = \frac{1}{2}\ln\frac{33}{17}$ units^2

13. Let $u = x + 3$, so $\frac{du}{dx} = 1$ and thus $du = dx$. When $x = 10$, $u = 13$, and when $x = 25$, $u = 28$.

Finding the area: $\int_{10}^{25} \frac{800}{\sqrt[3]{(x+3)^2}}dx = 800\int_{13}^{28} u^{-2/3}du = 800 \cdot 3u^{1/3}\Big|_{13}^{28} = 2{,}400\left(28^{1/3} - 13^{1/3}\right) \approx 1{,}644.61$ units^2

15. Finding the area: $\int_5^{20}\left(\frac{1}{2}x^2 - \frac{1}{3}x\right)dx = \frac{1}{2} \cdot \frac{1}{3}x^3 - \frac{1}{3} \cdot \frac{1}{2}x^2\Big|_5^{20} = \frac{1}{6}(20^3 - 20^2) - \frac{1}{6}(5^3 - 5^2) = \frac{7{,}600}{6} - \frac{100}{6} = 1{,}250$ units^2

17. Finding the area, making sure to use $-f(x)$ for the interval where $f(x)$ is negative:

$$\int_0^2 \left(x^2 - 7x + 10\right)dx - \int_2^5 \left(x^2 - 7x + 10\right)dx + \int_5^{10}\left(x^2 - 7x + 10\right)dx$$

$$= \left(\frac{1}{3}x^3 - \frac{7}{2}x^2 + 10x\right)\Big|_0^2 - \left(\frac{1}{3}x^3 - \frac{7}{2}x^2 + 10x\right)\Big|_2^5 + \left(\frac{1}{3}x^3 - \frac{7}{2}x^2 + 10x\right)\Big|_5^{10}$$

$$= \left[\left(\frac{8}{3} - 14 + 20\right) - 0\right] - \left[\left(\frac{125}{3} - \frac{175}{2} + 50\right) - \left(\frac{8}{3} - 14 + 20\right)\right] + \left[\left(\frac{1{,}000}{3} - \frac{700}{2} + 100\right) - \left(\frac{125}{3} - \frac{175}{2} + 50\right)\right]$$

$$= \frac{26}{3} - \left(-\frac{9}{2}\right) + \frac{475}{6}$$

$$= \frac{277}{3} \text{ units}^2$$

19. Finding the area, making sure to use $-f(x)$ for the interval where $f(x)$ is negative:

$$-\int_2^4 \frac{20}{x-5}dx + \int_6^8 \frac{20}{x-5}dx = \left(-20\ln|x-5|\right)\Big|_2^4 + \left(20\ln|x-5|\right)\Big|_6^8$$

$$= -20(\ln 1 - \ln 3) + 20(\ln 3 - \ln 1)$$

$$= 20\ln 3 + 20\ln 3$$

$$= 40\ln 3$$

$$\approx 43.94 \text{ units}^2$$

21. The area can be written as: $\int_a^b f(x)dx + \int_b^c f(x)dx$

23. The area can be written as: $\int_a^b f(x)\,dx - \int_a^c f(x)\,dx$

25. Let $u = 1 + 0.04x$, so $\dfrac{du}{dx} = 0.04$ and thus $\dfrac{1}{0.04}\,du = dx$. When $x = 0$, $u = 1$, and when $x = 400$, $u = 17$.

Finding the integral: $\displaystyle\int_0^{400} \dfrac{350}{1 + 0.04x}\,dx = \dfrac{350}{0.04}\int_1^{17}\dfrac{1}{u}\,du = 8{,}750\ln|u|\Big|_1^{17} = 8{,}750\ln 17 \approx \$24{,}790.62$

27. Let $u = 0.08t$, so $\dfrac{du}{dt} = 0.08$ and thus $\dfrac{1}{0.08}\,du = dt$. When $t = 5$, $u = 0.40$, and when $t = 8$, $u = 0.64$.

Finding the integral: $\displaystyle\int_5^8 400e^{0.08t}\,dt = \dfrac{400}{0.08}\int_{0.40}^{0.64}e^u\,du = 5{,}000e^u\Big|_{0.40}^{0.64} = 5{,}000\left(e^{0.64} - e^{0.40}\right) \approx \$2{,}023.28$

29. Let $u = 4t + 2$, so $\dfrac{du}{dt} = 4$ and thus $\dfrac{1}{4}\,du = dt$. When $t = 0$, $u = 2$, and when $t = 3$, $u = 14$.

Finding the integral: $\displaystyle\int_0^3 2{,}650(4t + 2)^{1/3}\,dt = \dfrac{2{,}650}{4}\int_2^{14}u^{1/3}\,du = \dfrac{1{,}325}{2}\cdot\dfrac{3}{4}u^{4/3}\Big|_2^{14} = \dfrac{3{,}975}{8}\left(14^{4/3} - 2^{4/3}\right) \approx \$15{,}513.51$

31. Finding the integral: $\displaystyle\int_0^8\left(-2.7t^2 + 26t\right)dt = -0.9t^3 + 13t^2\Big|_0^8 = -460.8 + 832 = 371.2$ units

33. Sketching the graph:

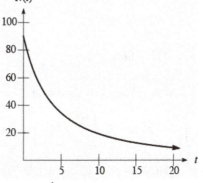

Let $u = 0.65t + 3.6$, so $\dfrac{du}{dt} = 0.65$ and thus $\dfrac{1}{0.65}\,du = dt$. When $t = 5$, $u = 6.85$, and when $t = 10$, $u = 10.1$.

Finding the integral:

$$\int_5^{10}\dfrac{615}{(0.65t + 3.6)^{3/2}}\,dt = \dfrac{615}{0.65}\int_{6.85}^{10.1}u^{-3/2}\,du = \dfrac{12{,}300}{13}\cdot\left(-2u^{-1/2}\right)\Big|_{6.85}^{10.1} = -\dfrac{24{,}600}{13}\left(\dfrac{1}{\sqrt{10.1}} - \dfrac{1}{\sqrt{6.85}}\right) \approx 127.6 \text{ gallons}$$

35. Finding the integral:

$$\int_0^4\left(243t^4 + 2{,}739t^3 - 31{,}797t^2 + 81{,}861t + 784{,}628\right)dt$$

$$= 48.6t^5 + 684.75t^4 - 10{,}599t^3 + 40{,}930.5t^2 + 784{,}628t\Big|_0^4$$

$$= 49{,}766.4 + 175{,}296 - 678{,}336 + 654{,}888 + 3{,}138{,}512$$

$$= \$3{,}340{,}126.4 \text{ million}$$

37. Evaluating the integrals:

$$\int_1^{100{,}000}\dfrac{1}{x}\,dx = \ln|x|\Big|_1^{100{,}000} = \ln 100{,}000 - 0 \approx 11.5129$$

$$\int_1^2\dfrac{1}{x}\,dx = \ln|x|\Big|_1^2 = \ln 2 - 0 \approx 0.6931$$

39. Let $u = x - 2$, so $\dfrac{du}{dx} = 1$ and thus $du = dx$. When $x = b$, $u = b - 2$, and when $x = 1$, $u = -1$.

Finding the integral: $\displaystyle\int_b^1 \frac{1}{(x-2)^3}\,dx = \int_{b-2}^{-1} u^{-3}\,du = -\frac{1}{2}u^{-2}\Big|_{b-2}^{-1} = -\frac{1}{2}\left(1 - \frac{1}{(b-2)^2}\right) = -\frac{1}{2} + \frac{1}{2(b-2)^2}$

41. Finding the limit:

$$100\lim_{b\to\infty}\left[\left(-2e^{-0.5b} + 1.25e^{-0.8b}\right) - \left(-2e^{-0.5(0)} + 1.25e^{-0.8(0)}\right)\right] = 100\lim_{b\to\infty}\left[\left(\frac{-2}{e^{0.5b}} + \frac{1.25}{e^{0.8b}}\right) - (-2 + 1.25)\right]$$
$$= 100(0 + 0 + 0.75)$$
$$= 75$$

5.5 Improper Integrals

1. Finding the integral: $\displaystyle\int_1^b \frac{1}{x^4}\,dx = -\frac{1}{3}x^{-3}\Big|_1^b = -\frac{1}{3}\left(\frac{1}{b^3} - 1\right) = \frac{1}{3} - \frac{1}{3b^3}$

Now taking limits: $\displaystyle\int_1^\infty \frac{1}{x^4}\,dx = \lim_{b\to\infty}\left(\frac{1}{3} - \frac{1}{3b^3}\right) = \frac{1}{3}$

3. Finding the integral: $\displaystyle\int_e^b \frac{3}{x}\,dx = 3\ln|x|\Big|_e^b = 3(\ln b - 1)$

Now taking limits: $\displaystyle\int_e^\infty \frac{3}{x}\,dx = \lim_{b\to\infty}\left[3(\ln b - 1)\right]$, which is divergent

5. Let $u = 3x + 4$, so $\dfrac{du}{dx} = 3$ and thus $\dfrac{1}{3}du = dx$. When $x = 0$, $u = 4$, and when $x = b$, $u = 3b + 4$.

Finding the integral: $\displaystyle\int_0^b \frac{1}{3x+4}\,dx = \frac{1}{3}\int_4^{3b+4} \frac{1}{u}\,du = \frac{1}{3}\ln|u|\Big|_4^{3b+4} = \frac{1}{3}\left(\ln(3b+4) - \ln 4\right)$

Now taking limits: $\displaystyle\int_0^\infty \frac{1}{3x+4}\,dx = \lim_{b\to\infty}\left[\frac{1}{3}\left(\ln(3b+4) - \ln 4\right)\right]$, which is divergent

7. Let $u = -4x$, so $\dfrac{du}{dx} = -4$ and thus $-\dfrac{1}{4}du = dx$. When $x = 0$, $u = 0$, and when $x = b$, $u = -4b$.

Finding the integral: $\displaystyle\int_0^b e^{-4x}\,dx = -\frac{1}{4}\int_0^{-4b} e^u\,du = -\frac{1}{4}e^u\Big|_0^{-4b} = -\frac{1}{4}\left(e^{-4b} - 1\right) = \frac{1}{4} - \frac{1}{4e^{4b}}$

Now taking limits: $\displaystyle\int_0^\infty e^{-4x}\,dx = \lim_{b\to\infty}\left(\frac{1}{4} - \frac{1}{4e^{4b}}\right) = \frac{1}{4}$

9. Let $u = 2x$, so $\dfrac{du}{dx} = 2$ and thus $\dfrac{1}{2}du = dx$. When $x = 0$, $u = 0$, and when $x = -b$, $u = -2b$.

Finding the integral: $\displaystyle\int_{-b}^0 e^{2x}\,dx = \frac{1}{2}\int_{-2b}^0 e^u\,du = \frac{1}{2}e^u\Big|_{-2b}^0 = \frac{1}{2}\left(1 - e^{-2b}\right) = \frac{1}{2} - \frac{1}{2e^{2b}}$

Now taking limits: $\displaystyle\int_{-\infty}^0 e^{2x}\,dx = \lim_{b\to\infty}\left(\frac{1}{2} - \frac{1}{2e^{2b}}\right) = \frac{1}{2}$

11. Let $u = x + 1$, so $\dfrac{du}{dx} = 1$ and thus $du = dx$. When $x = 0$, $u = 1$, and when $x = b$, $u = b + 1$.

Finding the integral: $\displaystyle\int_0^b \frac{1}{(x+1)^3}\,dx = \int_1^{b+1} u^{-3}\,du = -\frac{1}{2}u^{-2}\Big|_1^{b+1} = -\frac{1}{2}\left(\frac{1}{(b+1)^2} - 1\right) = \frac{1}{2} - \frac{1}{2(b+1)^2}$

Now taking limits: $\displaystyle\int_0^\infty \frac{1}{(x+1)^3}\,dx = \lim_{b\to\infty}\left[\frac{1}{2} - \frac{1}{2(b+1)^2}\right] = \frac{1}{2}$

13. Finding the integral: $\int_1^b \dfrac{6}{x^7}\,dx = -x^{-6}\Big|_1^b = -\left(\dfrac{1}{b^6}-1\right) = 1 - \dfrac{1}{b^6}$

Now taking limits: $\int_1^\infty \dfrac{6}{x^7}\,dx = \lim_{b\to\infty}\left(1 - \dfrac{1}{b^6}\right) = 1$

15. Let $u = -x^2$, so $\dfrac{du}{dx} = -2x$ and thus $-\dfrac{1}{2}\,du = x\,dx$. When $x = 0$, $u = 0$, and when $x = b$, $u = -b^2$.

Finding the integral: $\int_0^b xe^{-x^2}\,dx = -\dfrac{1}{2}\int_0^{-b^2} e^u\,du = -\dfrac{1}{2}e^u\Big|_0^{-b^2} = -\dfrac{1}{2}\left(e^{-b^2}-1\right) = \dfrac{1}{2} - \dfrac{1}{2e^{b^2}}$

Now taking limits: $\int_0^\infty xe^{-x^2}\,dx = \lim_{b\to\infty}\left(\dfrac{1}{2} - \dfrac{1}{2e^{b^2}}\right) = \dfrac{1}{2}$

17. Let $u = x^3 + 6$, so $\dfrac{du}{dx} = 3x^2$ and thus $2\,du = 6x^2\,dx$. When $x = 0$, $u = 6$, and when $x = b$, $u = b^3 + 6$.

Finding the integral: $\int_0^b \dfrac{6x^2}{x^3+6}\,dx = 2\int_6^{b^3+6} \dfrac{1}{u}\,du = 2\ln|u|\Big|_6^{b^3+6} = 2\left(\ln\left(b^3+6\right)-\ln 6\right)$

Now taking limits: $\int_0^\infty \dfrac{6x^2}{x^3+6}\,dx = \lim_{b\to\infty}\left[2\left(\ln\left(b^3+6\right)-\ln 6\right)\right]$, which is divergent

19. Let $u = \ln x$, so $\dfrac{du}{dx} = \dfrac{1}{x}$ and thus $du = \dfrac{1}{x}\,dx$. When $x = e$, $u = 1$, and when $x = b$, $u = \ln b$.

Finding the integral: $\int_e^b \dfrac{1}{x(\ln x)^2}\,dx = \int_1^{\ln b} \dfrac{1}{u^2}\,du = -u^{-1}\Big|_1^{\ln b} = -\left(\dfrac{1}{\ln b}-1\right) = 1 - \dfrac{1}{\ln b}$

Now taking limits: $\int_e^\infty \dfrac{1}{x(\ln x)^2}\,dx = \lim_{b\to\infty}\left(1 - \dfrac{1}{\ln b}\right) = 1$

21. Let $u = -ax$, so $\dfrac{du}{dx} = -a$ and thus $-du = a\,dx$. When $x = 0$, $u = 0$, and when $x = b$, $u = -ab$.

Finding the integral: $\int_0^b ae^{-ax}\,dx = -\int_0^{-ab} e^u\,du = -e^u\Big|_0^{-ab} = -\left(e^{-ab}-1\right) = 1 - \dfrac{1}{e^{ab}}$

Now taking limits: $\int_0^\infty ae^{-ax}\,dx = \lim_{b\to\infty}\left(1 - \dfrac{1}{e^{ab}}\right) = 1$

23. Note there is a discontinuity (undefined) at $x = 0$. Finding the integral:

$\int_{-1}^1 x^{-4/5}\,dx = \lim_{b\to 0^-}\int_{-1}^b x^{-4/5}\,dx + \lim_{a\to 0^+}\int_a^1 x^{-4/5}\,dx$

$= \lim_{b\to 0^-}\left[5x^{1/5}\Big|_{-1}^b\right] + \lim_{a\to 0^+}\left[5x^{1/5}\Big|_a^1\right]$

$= \lim_{b\to 0^-}\left(5b^{1/5}+5\right) + \lim_{a\to 0^+}\left(5 - 5a^{1/5}\right)$

$= 5 + 5$

$= 10$

25. A dose taken orally will only be 85% as effective as the same dose taken intravenously.

27. The integral represents the total amount of money received from the rental property from now into the future.

29. The integral represents the total amount of energy conserved starting 1.5 years from now into the future.

31. The integral represents the total amount of improvement the company can expect from now into the future.

33. Let $u = -0.04t$, so $\dfrac{du}{dt} = -0.04$ and thus $-\dfrac{1}{0.04}du = dt$. When $t = 0$, $u = 0$, and when $t = b$, $u = -0.04b$.

Finding the integral: $\displaystyle\int_0^b 1{,}200e^{-0.04t}dt = \dfrac{1{,}200}{-0.04}\int_0^{-0.04b} e^u\,du = -30{,}000\,e^u\Big|_0^{-0.04b} = -30{,}000\left(e^{-0.04b}-1\right) = 30{,}000\left(1-\dfrac{1}{e^{0.04b}}\right)$

Now taking limits: $\displaystyle\int_0^\infty 1{,}200e^{-0.04t}dt = \lim_{b\to\infty}\left[30{,}000\left(1-\dfrac{1}{e^{0.04b}}\right)\right] = 30{,}000$ tons

35. Let $u = -15t$, so $\dfrac{du}{dt} = -15$ and thus $-du = 15\,dt$. When $t = \dfrac{1}{12}$, $u = -\dfrac{5}{4}$, and when $t = b$, $u = -15b$.

Finding the integral: $\displaystyle\int_{1/12}^b 15e^{-15t}dt = -\int_{-5/4}^{-15b} e^u\,du = -e^u\Big|_{-5/4}^{-15b} = -\left(e^{-15b}-e^{-5/4}\right) = e^{-5/4}-\dfrac{1}{e^{15b}}$

Now taking limits: $\displaystyle\int_{1/12}^\infty 15e^{-15t}dt = \lim_{b\to\infty}\left(e^{-5/4}-\dfrac{1}{e^{15b}}\right) = \dfrac{1}{e^{5/4}} \approx 0.2865$

37. Let $u = -0.26t$, so $\dfrac{du}{dt} = -0.26$ and thus $-\dfrac{1}{0.26}du = dt$. When $t = 0$, $u = 0$, and when $t = b$, $u = -0.26b$.

Finding the integral: $\displaystyle\int_0^b 400.4e^{-0.26t}dt = \dfrac{400.4}{-0.26}\int_0^{-0.26b} e^u\,du = -1{,}540\,e^u\Big|_0^{-0.26b} = -1{,}540\left(e^{-0.26b}-1\right) = 1{,}540\left(1-\dfrac{1}{e^{0.26b}}\right)$

Now taking limits: $\displaystyle\int_0^\infty 400.4e^{-0.26t}dt = \lim_{b\to\infty}\left[1{,}540\left(1-\dfrac{1}{e^{0.26b}}\right)\right] = 1{,}540$ units

39. The capital value is given by the integral: $\displaystyle\int_0^\infty 12{,}000e^{0.05t}e^{-0.10t}dt = \int_0^\infty 12{,}000e^{-0.05t}dt$

Let $u = -0.05t$, so $\dfrac{du}{dt} = -0.05$ and thus $-\dfrac{1}{0.05}du = dt$. When $t = 0$, $u = 0$, and when $t = b$, $u = -0.05b$.

Finding the integral:

$$\int_0^b 12{,}000e^{-0.05t}dt = \dfrac{12{,}000}{-0.05}\int_0^{-0.05b} e^u\,du = -240{,}000\,e^u\Big|_0^{-0.05b} = -240{,}000\left(e^{-0.05b}-1\right) = 240{,}000\left(1-\dfrac{1}{e^{0.05b}}\right)$$

Now taking limits: $\displaystyle\int_0^\infty 12{,}000e^{-0.05t}dt = \lim_{b\to\infty}\left[240{,}000\left(1-\dfrac{1}{e^{0.05b}}\right)\right] = \$240{,}000$

41. **a.** Let $u = -rt$, so $\dfrac{du}{dt} = -r$ and thus $-\dfrac{1}{r}du = dt$. When $t = 0$, $u = 0$, and when $t = b$, $u = -rb$.

Finding the integral: $P\displaystyle\int_0^b e^{-rt}dt = -\dfrac{P}{r}\int_0^{-rb} e^u\,du = -\dfrac{P}{r}e^u\Big|_0^{-rb} = -\dfrac{P}{r}\left(e^{-rb}-1\right) = \dfrac{P}{r}\left(1-\dfrac{1}{e^{rb}}\right)$

Now taking limits: $P\displaystyle\int_0^\infty e^{-rt}dt = \lim_{b\to\infty}\left[\dfrac{P}{r}\left(1-\dfrac{1}{e^{rb}}\right)\right] = \dfrac{P}{r}$

b. Using $P = \$30{,}000$ and $r = 0.05$, the present value is: $\dfrac{P}{r} = \dfrac{\$30{,}000}{0.05} = \$600{,}000$

43. For the first integral, make two u-substitutions to obtain:

$$\int_0^b 300\left(e^{-0.6t}-e^{-1.5t}\right)dt = \dfrac{300}{-0.6}\int_0^{-0.6b} e^u\,du - \dfrac{300}{-1.5}\int_0^{-1.5b} e^u\,du$$

$$= -500\,e^u\Big|_0^{-0.6b} + 200\,e^u\Big|_0^{-1.5b}$$

$$= -500\left(e^{-0.6b}-1\right) + 200\left(e^{-1.5b}-1\right)$$

$$= 100\left(-5e^{-0.6b}+5+2e^{-1.5b}-2\right)$$

$$= 100\left(3-\dfrac{5}{e^{0.6b}}+\dfrac{2}{e^{1.5b}}\right)$$

Now taking limits: $\int_0^\infty C_{po}(t)\,dt = \lim_{b\to\infty}\left[100\left(3-\dfrac{5}{e^{0.6b}}+\dfrac{2}{e^{1.5b}}\right)\right]=300$

Finding the second integral: $\int_0^b 300\left(e^{-0.6t}\right)dt = \dfrac{300}{-0.6}\int_0^{-0.6b} e^u\,du = -500\,e^u\Big|_0^{-0.6b} = -500\left(e^{-0.6b}-1\right)=500\left(1-\dfrac{1}{e^{0.6b}}\right)$

Now taking limits: $\int_0^\infty C_{iv}(t)\,dt = \lim_{b\to\infty}\left[500\left(1-\dfrac{1}{e^{0.6b}}\right)\right]=500$

The bioavailability is given by: $F=\dfrac{\int_0^\infty C_{po}(t)\,dt}{\int_0^\infty C_{iv}(t)\,dt}=\dfrac{300}{500}=\dfrac{3}{5}=0.6$

Taken orally, the drug is 60% as effective as a dose taken internally.

45. Since there is a discontinuity at $x=0$, we need to evaluate this as two separate improper integrals:

$$\int_{-\infty}^\infty \frac{1}{x}\,dx = \int_{-\infty}^0 \frac{1}{x}\,dx + \int_0^\infty \frac{1}{x}\,dx = \lim_{b\to\infty}\lim_{a\to0^-}\int_{-b}^a \frac{1}{x}\,dx + \lim_{b\to\infty}\lim_{a\to0^+}\int_a^b \frac{1}{x}\,dx$$

Evaluating the first integral: $\int_{-b}^a \dfrac{1}{x}\,dx = \ln|x|\Big|_{-b}^a = \ln|a|-\ln|b|$

Since both of these values diverge, the integral is divergent.

47. Let $u=-0.22t$, so $\dfrac{du}{dt}=-0.22$ and thus $-\dfrac{1}{0.22}\,du=dt$. When $t=0$, $u=0$, and when $t=b$, $u=-0.22b$.

Finding the integral: $\int_0^b 36e^{-0.22t}\,dt = \dfrac{36}{-0.22}\int_0^{-0.22b} e^u\,du = -\dfrac{1,800}{11}e^u\Big|_0^{-0.22b} = -\dfrac{1,800}{11}\left(e^{-0.22b}-1\right) = \dfrac{1,800}{11}\left(1-\dfrac{1}{e^{0.22b}}\right)$

Now taking limits: $\int_0^\infty 36e^{-0.22t}\,dt = \lim_{b\to\infty}\left[\dfrac{1,800}{11}\left(1-\dfrac{1}{e^{0.22b}}\right)\right]\approx\$163{,}636{,}364$

The total income the company can expect is \$163,636,364.

49. Let $u=-x^2$, so $\dfrac{du}{dx}=-2x$ and thus $-du=2x\,dx$. When $x=b$, $u=-b^2$, and when $x=a$, $u=-a^2$.

Finding the integral: $\int_b^a 2xe^{-x^2}\,dx = -\int_{-b^2}^{-a^2} e^u\,du = -e^u\Big|_{-b^2}^{-a^2} = -\left(e^{-a^2}-e^{-b^2}\right) = \dfrac{1}{e^{b^2}}-\dfrac{1}{e^{a^2}}$

Now taking limits: $\int_{-\infty}^\infty 2xe^{-x^2}\,dx = \lim_{\substack{b\to\infty\\a\to\infty}}\left[\dfrac{1}{e^{b^2}}-\dfrac{1}{e^{a^2}}\right]=0$

The graph indicates the total (net) area under the curve is 0:

51. Writing the expression: $uv-\int v\cdot du = (\ln x)\left(\dfrac{1}{5}x^5\right)-\int \dfrac{1}{5}x^5\cdot\dfrac{1}{x}\,dx = \dfrac{1}{5}x^5\ln x-\int\dfrac{1}{5}x^4\,dx$

5.6 Integration by Parts

1. Making the substitutions:

$$u = x \qquad\qquad dv = e^x dx$$

$$du = dx \qquad\qquad v = e^x$$

Integrating by parts: $\int xe^x\,dx = uv - \int v \cdot du = xe^x - \int e^x\,dx = xe^x - e^x + C = e^x(x-1) + C$

3. Making the substitutions:

$$u = x \qquad\qquad dv = e^{-8x} dx$$

$$du = dx \qquad\qquad v = -\frac{1}{8}e^{-8x}$$

Integrating by parts:

$$\int xe^{-8x}\,dx = uv - \int v \cdot du = x \cdot \left(-\frac{1}{8}e^{-8x}\right) + \frac{1}{8}\int e^{-8x}\,dx = -\frac{1}{8}xe^{-8x} - \frac{1}{64}e^{-8x} + C = -\frac{1}{64}e^{-8x}(8x+1) + C$$

5. Making the substitutions:

$$u = x^3 \qquad\qquad dv = e^{3x} dx$$

$$du = 3x^2 dx \qquad\qquad v = \frac{1}{3}e^{3x}$$

Integrating by parts: $\int x^3 e^{3x}\,dx = uv - \int v \cdot du = x^3 \cdot \left(\frac{1}{3}e^{3x}\right) - \int 3x^2 \cdot \frac{1}{3}e^{3x}\,dx = \frac{1}{3}x^3 e^{3x} - \int x^2 e^{3x}\,dx$

For the second integral, we must use parts again. Making the substitutions:

$$u = x^2 \qquad\qquad dv = e^{3x} dx$$

$$du = 2x dx \qquad\qquad v = \frac{1}{3}e^{3x}$$

Integrating by parts: $\int x^2 e^{3x}\,dx = uv - \int v \cdot du = x^2 \cdot \left(\frac{1}{3}e^{3x}\right) - \int 2xe^{3x}\,dx = \frac{1}{3}x^2 e^{3x} - \frac{2}{3}\int xe^{3x}\,dx$

So the original integral becomes:

$$\int x^3 e^{3x}\,dx = \frac{1}{3}x^3 e^{3x} - \left(\frac{1}{3}x^2 e^{3x} - \frac{2}{3}\int xe^{3x}\,dx\right) = \frac{1}{3}x^3 e^{3x} - \frac{1}{3}x^2 e^{3x} + \frac{2}{3}\int xe^{3x}\,dx$$

For the last integral, we must use parts again. Making the substitutions:

$$u = x \qquad\qquad dv = e^{3x} dx$$

$$du = dx \qquad\qquad v = \frac{1}{3}e^{3x}$$

Integrating by parts: $\int xe^{3x}\,dx = uv - \int v \cdot du = x \cdot \left(\frac{1}{3}e^{3x}\right) - \frac{1}{3}\int e^{3x}\,dx = \frac{1}{3}xe^{3x} - \frac{1}{9}e^{3x}$

Making the final substitutions:

$$\int x^3 e^{3x}\,dx = \frac{1}{3}x^3 e^{3x} - \frac{1}{3}x^2 e^{3x} + \frac{2}{3}\int xe^{3x}\,dx$$

$$= \frac{1}{3}x^3 e^{3x} - \frac{1}{3}x^2 e^{3x} + \frac{2}{3}\left(\frac{1}{3}xe^{3x} - \frac{1}{9}e^{3x}\right) + C$$

$$= \frac{1}{3}x^3 e^{3x} - \frac{1}{3}x^2 e^{3x} + \frac{2}{9}xe^{3x} - \frac{2}{27}e^{3x} + C$$

$$= \frac{1}{27}e^{3x}\left(9x^3 - 9x^2 + 6x - 2\right) + C$$

7. Making the substitutions:
$$u = \ln\left(x^3\right) \qquad\qquad dv = dx$$
$$du = \frac{1}{x^3} \cdot 3x^2 dx = \frac{3}{x} dx \qquad\qquad v = x$$

Integrating by parts:
$$\int \ln\left(x^3\right) dx = uv - \int v \cdot du = x\ln\left(x^3\right) - \int x \cdot \frac{3}{x} dx = x\ln\left(x^3\right) - 3\int dx = x\ln\left(x^3\right) - 3x + C = x\left(\ln\left(x^3\right) - 3\right) + C$$

9. Making the substitutions:
$$u = \left(\ln x\right)^3 \qquad\qquad dv = dx$$
$$du = 3\left(\ln x\right)^2 \cdot \frac{1}{x} dx \qquad\qquad v = x$$

Integrating by parts: $\int \left(\ln x\right)^3 dx = uv - \int v \cdot du = x\left(\ln x\right)^3 - \int x \cdot 3\left(\ln x\right)^2 \cdot \frac{1}{x} dx = x\left(\ln x\right)^3 - 3\int \left(\ln x\right)^2 dx$

For the second integral, we must integrate by parts again. Making the substitutions:
$$u = \left(\ln x\right)^2 \qquad\qquad dv = dx$$
$$du = 2\left(\ln x\right) \cdot \frac{1}{x} dx \qquad\qquad v = x$$

Integrating by parts: $\int \left(\ln x\right)^2 dx = uv - \int v \cdot du = x\left(\ln x\right)^2 - \int x \cdot 2\left(\ln x\right) \cdot \frac{1}{x} dx = x\left(\ln x\right)^2 - 2\int \ln x \, dx$

Substituting into the original integral: $\int \left(\ln x\right)^3 dx = x\left(\ln x\right)^3 - 3\left(x\left(\ln x\right)^2 - 2\int \ln x \, dx\right) = x\left(\ln x\right)^3 - 3x\left(\ln x\right)^2 + 6\int \ln x \, dx$

For the last integral, we must integrate by parts again. Making the substitutions:
$$u = \ln x \qquad\qquad dv = dx$$
$$du = \frac{1}{x} dx \qquad\qquad v = x$$

Integrating by parts: $\int \ln x \, dx = uv - \int v \cdot du = x\ln x - \int x \cdot \frac{1}{x} dx = x\ln x - \int dx = x\ln x - x + C$

Substituting into the original integral:
$$\int \left(\ln x\right)^3 dx = x\left(\ln x\right)^3 - 3x\left(\ln x\right)^2 + 6\left(x\ln x - x\right) + C$$
$$= x\left(\ln x\right)^3 - 3x\left(\ln x\right)^2 + 6x\ln x - 6x + C$$
$$= x\left(\left(\ln x\right)^3 - 3\left(\ln x\right)^2 + 6\ln x - 6\right) + C$$

11. This can be integrated using a regular u-substitution. Let $u = \ln x$, so $du = \frac{1}{x} dx$. Substituting:
$$\int \frac{\ln x}{x} dx = \int u \, du = \frac{1}{2} u^2 + C = \frac{1}{2}\left(\ln x\right)^2 + C$$

13. Making the substitutions:

$$u = (\ln x)^3 \qquad\qquad dv = x\,dx$$

$$du = 3(\ln x)^2 \cdot \frac{1}{x}\,dx \qquad\qquad v = \frac{1}{2}x^2$$

Integrating by parts: $\displaystyle\int x(\ln x)^3\,dx = uv - \int v\cdot du = \frac{1}{2}x^2(\ln x)^3 - \int \frac{1}{2}x^2 \cdot 3(\ln x)^2 \cdot \frac{1}{x}\,dx = \frac{1}{2}x^2(\ln x)^3 - \frac{3}{2}\int x(\ln x)^2\,dx$

For the second integral, we must integrate by parts again. Making the substitutions:

$$u = (\ln x)^2 \qquad\qquad dv = x\,dx$$

$$du = 2(\ln x)\cdot \frac{1}{x}\,dx \qquad\qquad v = \frac{1}{2}x^2$$

Integrating by parts: $\displaystyle\int x(\ln x)^2\,dx = uv - \int v\cdot du = \frac{1}{2}x^2(\ln x)^2 - \int \frac{1}{2}x^2 \cdot 2(\ln x)\cdot \frac{1}{x}\,dx = \frac{1}{2}x^2(\ln x)^2 - \int x\ln x\,dx$

Substituting into the original integral:

$$\int x(\ln x)^3\,dx = \frac{1}{2}x^2(\ln x)^3 - \frac{3}{2}\left(\frac{1}{2}x^2(\ln x)^2 - \int x\ln x\,dx\right) = \frac{1}{2}x^2(\ln x)^3 - \frac{3}{4}x^2(\ln x)^2 + \frac{3}{2}\int x\ln x\,dx$$

For the last integral, we must integrate by parts again. Making the substitutions:

$$u = \ln x \qquad\qquad dv = x\,dx$$

$$du = \frac{1}{x}\,dx \qquad\qquad v = \frac{1}{2}x^2$$

Integrating by parts:

$$\int \ln x\,dx = uv - \int v\cdot du = \frac{1}{2}x^2\ln x - \int \frac{1}{2}x^2 \cdot \frac{1}{x}\,dx = \frac{1}{2}x^2\ln x - \frac{1}{2}\int x\,dx = \frac{1}{2}x^2\ln x - \frac{1}{4}x^2 + C$$

Substituting into the original integral:

$$\int x(\ln x)^3\,dx = \frac{1}{2}x^2(\ln x)^3 - \frac{3}{4}x^2(\ln x)^2 + \frac{3}{2}\left(\frac{1}{2}x^2\ln x - \frac{1}{4}x^2\right) + C$$

$$= \frac{1}{2}x^2(\ln x)^3 - \frac{3}{4}x^2(\ln x)^2 + \frac{3}{4}x^2\ln x - \frac{3}{8}x^2 + C$$

$$= \frac{1}{8}x^2\left(4(\ln x)^3 - 6(\ln x)^2 + 6\ln x - 3\right) + C$$

15. Making the substitutions:

$$u = \ln(7x-4) \qquad\qquad dv = dx$$

$$du = \frac{1}{7x-4}\cdot 7\,dx = \frac{7}{7x-4}\,dx \qquad\qquad v = x$$

Integrating by parts: $\displaystyle\int \ln(7x-4)\,dx = uv - \int v\cdot du = x\ln(7x-4) - \int \frac{7x}{7x-4}\,dx$

For this second integral, let $u = 7x-4$, so $du = 7\,dx$ and $7x = u+4$. Substituting:

$$\int \frac{7x}{7x-4}\,dx = \frac{1}{7}\int \frac{u+4}{u}\,du = \frac{1}{7}\int \left(1 + \frac{4}{u}\right)\,du = \frac{1}{7}\left(u + 4\ln|u|\right) = \frac{1}{7}\left(7x-4 + 4\ln|7x-4|\right)$$

Substituting:

$$\int \ln(7x-4)\,dx = x\ln(7x-4) - \left(\frac{1}{7}\left(7x-4 + 4\ln|7x-4|\right)\right) + C$$

$$= \frac{1}{7}\left[7x\ln(7x-4) - (7x-4) - 4\ln|7x-4|\right] + C$$

$$= \frac{1}{7}\left[(7x-4)\ln(7x-4) - (7x-4)\right] + C$$

$$= \frac{1}{7}(7x-4)\left(\ln(7x-4) - 1\right) + C$$

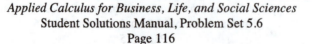

17. This can be integrated using a regular u-substitution. Let $u = 1 - x$, so $du = -dx$ and $x = 1 - u$. Substituting:

$$\int x(1-x)^{3/2}\,dx = -\int (1-u)u^{3/2}\,du$$
$$= -\int \left(u^{3/2} - u^{5/2}\right)du$$
$$= \int \left(u^{5/2} - u^{3/2}\right)du$$
$$= \frac{2}{7}u^{7/2} - \frac{2}{5}u^{5/2} + C$$
$$= \frac{2}{35}u^{5/2}(5u - 7) + C$$
$$= \frac{2}{35}(1-x)^{5/2}\left(5(1-x) - 7\right) + C$$
$$= \frac{2}{35}(1-x)^{5/2}(-5x - 2) + C$$
$$= -\frac{2}{35}(1-x)^{5/2}(5x + 2) + C$$

19. Making the substitutions:

$$u = 2x + 1 \qquad dv = e^{x+3}\,dx$$
$$du = 2dx \qquad v = e^{x+3}$$

Integrating by parts:

$$\int e^{x+3}(2x+1)\,dx = uv - \int v \cdot du = (2x+1)e^{x+3} - 2\int e^{x+3}\,dx = (2x+1)e^{x+3} - 2e^{x+3} + C = (2x-1)e^{x+3} + C$$

21. This integral is easiest to perform if we first make a substitution, then use parts. Let $y = 2x + 1$, so $dy = 2dx$ and $\frac{1}{2}dy = dx$. Substituting: $\int (2x+1)^2 \ln(2x+1)\,dx = \frac{1}{2}\int y^2 \ln y\,dy$

Now making the substitutions:

$$u = \ln y \qquad dv = y^2\,dy$$
$$du = \frac{1}{y}dy \qquad v = \frac{1}{3}y^3$$

Integrating by parts:

$$\int (2x+1)^2 \ln(2x+1)\,dx = \frac{1}{2}\int y^2 \ln y\,dy$$
$$= \frac{1}{2}uv - \frac{1}{2}\int v \cdot du$$
$$= \frac{1}{2}\left(\frac{1}{3}y^3 \ln y\right) - \frac{1}{2}\int \frac{1}{3}y^3 \cdot \frac{1}{y}dy$$
$$= \frac{1}{6}y^3 \ln y - \frac{1}{6}\int y^2\,dy$$
$$= \frac{1}{6}y^3 \ln y - \frac{1}{18}y^3 + C$$
$$= \frac{1}{18}y^3(3\ln y - 1) + C$$
$$= \frac{1}{18}(2x+1)^3\left(3\ln(2x+1) - 1\right) + C$$

23. Writing the expression: $\int 48xe^{0.16x}\,dx = uv - \int v \cdot du = 48x \cdot 6.25e^{0.16x} - 6.25\int e^{0.16x} \cdot 48\,dx = 300xe^{0.16x} - 300\int e^{0.16x}\,dx$

25. Writing the expression: $\int x^3 e^{-6x}\,dx = uv - \int v \cdot du = x^3 \cdot \left(-\frac{1}{6}e^{-6x}\right) + \frac{1}{6}\int e^{-6x} \cdot 3x^2\,dx = -\frac{1}{6}x^3 e^{-6x} + \frac{1}{2}\int x^2 e^{-6x}\,dx$

27. Writing the expression: $\int 6x^5 \ln 6x \, dx = uv - \int v \cdot du = \ln 6x \cdot \left(x^6\right) - \int x^6 \cdot \frac{1}{x} dx = x^6 \ln 6x - \int x^5 \, dx$

29. Making the substitutions:

$$u = x^2 \qquad\qquad\qquad dv = e^{-2x} dx$$

$$du = 2x dx \qquad\qquad\qquad v = -\frac{1}{2} e^{-2x}$$

Writing out the integrating by parts: $\int x^2 e^{-2x} \, dx = uv - \int v \cdot du = x^2 \left(-\frac{1}{2} e^{-2x}\right) + \frac{1}{2} \int e^{-2x}(2x) dx = -\frac{1}{2} x^2 e^{-2x} + \int x e^{-2x} \, dx$

31. Making the substitutions:

$$u = \ln 8x \qquad\qquad\qquad dv = 8x^7 dx$$

$$du = \frac{1}{8x} \cdot 8 dx = \frac{1}{x} dx \qquad v = x^8$$

Writing out the integrating by parts: $\int 8x^7 \ln 7x \, dx = uv - \int v \cdot du = x^8 \ln 8x - \int x^8 \cdot \frac{1}{x} dx = x^8 \ln 8x - \int x^7 \, dx$

33. Making the substitutions:

$$u = \ln(x+1) \qquad\qquad dv = x dx$$

$$du = \frac{1}{x+1} dx \qquad\qquad v = \frac{1}{2} x^2$$

Integrating by parts: $C(x) = \int x \ln(x+1) dx = uv - \int v \cdot du = \frac{1}{2} x^2 \ln(x+1) - \frac{1}{2} \int \frac{x^2}{x+1} dx$

For this second integral, let $u = x+1$, so $du = dx$ and $x = u-1$. Substituting:

$$\int \frac{x^2}{x+1} dx = \int \frac{(u-1)^2}{u} du = \int \frac{u^2 - 2u + 1}{u} du = \int \left(u - 2 + \frac{1}{u}\right) du = \frac{1}{2} u^2 - 2u + \ln u = \frac{1}{2}(x+1)^2 - 2(x+1) + \ln(x+1)$$

Substituting:

$$C(x) = \frac{1}{2} x^2 \ln(x+1) - \frac{1}{2}\left(\frac{1}{2}(x+1)^2 - 2(x+1) + \ln(x+1)\right) + C$$

$$= \frac{1}{2} x^2 \ln(x+1) - \frac{1}{4}(x+1)^2 + x - \frac{1}{2} \ln(x+1) + C$$

$$= \left(\frac{1}{2} x^2 - \frac{1}{2}\right) \ln(x+1) - \frac{1}{4}(x+1)^2 + x + C$$

$$= \frac{1}{2}(x^2 - 1) \ln(x+1) - \frac{1}{4}(x+1)^2 + x + C$$

Evaluating C when $C(0) = 0$:

$$C(0) = 0$$

$$\frac{1}{2}(-1) \ln(1) - \frac{1}{4}(1)^2 + 0 + C = 0$$

$$-\frac{1}{4} + C = 0$$

$$C = \frac{1}{4}$$

So the function is given by:
$$C(x) = \frac{1}{2}\left(x^2-1\right)\ln(x+1) - \frac{1}{4}(x+1)^2 + x + \frac{1}{4}$$
$$= \frac{1}{2}\left(x^2-1\right)\ln(x+1) - \frac{1}{4}x^2 - \frac{1}{2}x - \frac{1}{4} + x + \frac{1}{4}$$
$$= \frac{1}{2}\left(x^2-1\right)\ln(x+1) - \frac{1}{4}x^2 + \frac{1}{2}x$$
$$= \frac{1}{4}\left[2\left(x^2-1\right)\ln(x+1) - x^2 + 2x\right]$$

35. First write the integral in two parts:
$$r(t) = \int R(t)\,dt = \int\left(12{,}500 + 120.5t^2 e^{-t/2}\right)dt = 12{,}500\int dt + 120.5\int t^2 e^{-t/2}\,dt$$
Making the substitutions:

$u = t^2$ $\qquad\qquad$ $dv = e^{-t/2}dt$

$du = 2t\,dt$ $\qquad\qquad$ $v = -2e^{-t/2}$

Integrating by parts:
$$r(t) = 12{,}500t + 120.5\left(uv - \int v\cdot du\right) = 12{,}500t - 241t^2 e^{-t/2} + 482\int te^{-t/2}\,dt$$
Making the substitutions:

$u = t$ $\qquad\qquad$ $dv = e^{-t/2}dt$

$du = dt$ $\qquad\qquad$ $v = -2e^{-t/2}$

Integrating by parts again:
$$r(t) = 12{,}500t - 241t^2 e^{-t/2} + 482\left(uv - \int v\cdot du\right)$$
$$= 12{,}500t - 241t^2 e^{-t/2} + 482\left(-2te^{-t/2} + \int 2e^{-t/2}\,dt\right)$$
$$= 12{,}500t - 241t^2 e^{-t/2} + 482\left(-2te^{-t/2} - 4e^{-t/2}\right) + C$$
$$= 12{,}500t - 241e^{-t/2}\left(t^2 + 4t + 8\right) + C$$
Evaluating C when $r(0) = 12{,}500$:
$$r(0) = 12{,}500$$
$$12{,}500(0) - 241e^{-0/2}\left(0^2 + 4(0) + 8\right) + C = 12{,}500$$
$$-1{,}928 + C = 12{,}500$$
$$C = 14{,}428$$
So the function is given by: $r(t) = 12{,}500t - 241e^{-t/2}\left(t^2 + 4t + 8\right) + 14{,}428$

37. Making the substitutions:

$u = 8t$ $\qquad\qquad$ $dv = e^{-0.8t}dt$

$du = 8\,dt$ $\qquad\qquad$ $v = -1.25e^{-0.8t}$

Integrating by parts: $K(t) = uv - \int v\cdot du = -10te^{-0.8t} + 10\int e^{-0.8t}\,dt = -10te^{-0.8t} - 12.5e^{-0.8t} + C$

Evaluating C when $f(0) = 0$:
$$f(0) = 0$$
$$-10(0)e^{-0.8(0)} - 12.5e^{-0.8(0)} + C = 0$$
$$-12.5 + C = 0$$
$$C = 12.5$$
So the function is given by: $K(t) = -10te^{-0.8t} - 12.5e^{-0.8t} + 12.5 = -e^{-0.8t}\left(10t + 12.5\right) + 12.5$

39. Making the substitutions:

$$u = \ln(6x) \qquad\qquad dv = x\,dx$$

$$du = \frac{1}{6x} \cdot 6\,dx = \frac{1}{x}\,dx \qquad v = \frac{1}{2}x^2$$

Integrating by parts:

$$\int x\ln(6x)\,dx = uv - \int v \cdot du$$

$$= \frac{1}{2}x^2\ln(6x) - \frac{1}{2}\int x^2 \cdot \frac{1}{x}\,dx$$

$$= \frac{1}{2}x^2\ln(6x) - \frac{1}{2}\int x\,dx$$

$$= \frac{1}{2}x^2\ln(6x) - \frac{1}{2} \cdot \frac{1}{2}x^2 + C$$

$$= \frac{1}{4}x^2\big(2\ln(6x) - 1\big) + C$$

41. Using Wolfram|Alpha to find the integrals:

$$\int \frac{\ln x}{x^2}\,dx = -\frac{\ln x + 1}{x} + C$$

$$\int \frac{\ln x}{x^3}\,dx = -\frac{2\ln x + 1}{4x^2} + C$$

$$\int \frac{\ln x}{x^4}\,dx = -\frac{3\ln x + 1}{9x^3} + C$$

$$\int \frac{\ln x}{x^5}\,dx = -\frac{4\ln x + 1}{16x^4} + C$$

$$\int \frac{\ln x}{x^6}\,dx = -\frac{5\ln x + 1}{25x^5} + C$$

$$\int \frac{\ln x}{x^n}\,dx = -\frac{(n-1)\ln x + 1}{(n-1)^2 x^{n-1}} + C$$

43. Evaluating the integral: $\int\big(-x^2 + 10x - 16\big)dx = -\frac{1}{3}x^3 + 5x^2 - 16x + C$

45. Evaluating the integral: $\int\big[(-x^2 + 4x + 63) - (x^2 + 12x + 39)\big]dx = \int\big(-2x^2 - 8x + 24\big)dx = -\frac{2}{3}x^3 - 4x^2 + 24x + C$

47. Finding where the curves intersect:

$$x^3 - 10x + 25 = 6x + 25$$

$$x^3 - 16x = 0$$

$$x\big(x^2 - 16\big) = 0$$

$$x(x+4)(x-4) = 0$$

$$x = -4, 0, 4$$

Chapter 5 Test

1. If the air temperature increases by $1°$, the efficiency of the pup will decrease by 3.2%. When the surrounding air temperature is $10°$, the efficiency of the pump is 64%.

2. Evaluating the integral: $\displaystyle\int \frac{16}{x}dx = 16\ln|x| + C$

3. Evaluating the integral: $\displaystyle\int\left(3x^2 - 5x + 2\right)dx = x^3 - \frac{5}{2}x^2 + 2x + C$

4. Evaluating the integral: $\displaystyle\int 4\sqrt[5]{x^3}\,dx = \int 4x^{3/5}\,dx = 4\cdot\frac{5}{8}x^{8/5} + C = \frac{5}{2}x^{8/5} + C$

5. Evaluating the integral: $\displaystyle\int 40e^{-0.02x}\,dx = \frac{40}{-0.02}e^{-0.02x} + C = -2{,}000e^{-0.02x} + C$

6. Evaluating: $V'(5) = -12{,}480e^{-0.24(5)} \approx -3{,}758.90$

    After 5 days of the campaign, the sales volume is decreasing by $\$3{,}758.90$ per day.

    Finding the integral:
    $$V(t) = \int -12{,}480e^{-0.24t}\,dt = \frac{-12{,}480}{-0.24}e^{-0.24t} + C = 52{,}000e^{-0.24t} + C$$

    Since $V(0) = 104{,}000$, we can evaluate the constant:
    $$V(0) = 52{,}000e^{-0.24(0)} + C$$
    $$104{,}000 = 52{,}000 + C$$
    $$C = 52{,}000$$

    So $V(t) = 52{,}000e^{-0.24t} + 52{,}000$.

    Evaluating the function: $V(5) = 52{,}000e^{-0.24(5)} + 52{,}000 \approx 67{,}662.10$

    After 5 days of the campaign, the sales volume is $\$67{,}662.10$.

7. Let $u = 2x^2 - 5$, so $\dfrac{du}{dx} = 4x$ and thus $\dfrac{1}{4}du = xdx$. Finding the integral:
    $$\int x\left(2x^2 - 5\right)^{2/3}dx = \frac{1}{4}\int u^{2/3}\,du = \frac{1}{4}\cdot\frac{3}{5}u^{5/3} + C = \frac{3}{20}\left(2x^2 - 5\right)^{5/3} + C$$

8. Let $u = x^2 + 1$, so $\dfrac{du}{dx} = 2x$ and thus $\dfrac{1}{2}du = xdx$. Finding the integral:
    $$\int \frac{x}{x^2+1}dx = \frac{1}{2}\int\frac{1}{u}du = \frac{1}{2}\ln|u| + C = \frac{1}{2}\ln\left(x^2 + 1\right) + C$$

9. Let $u = x^2 + 4x - 1$, so $\dfrac{du}{dx} = 2x + 4$ and thus $du = 2(x+2)dx$. Finding the integral:
    $$\int 2(x+2)\sqrt{x^2 + 4x - 1}\,dx = \int u^{1/2}\,du = \frac{2}{3}u^{3/2} + C = \frac{2}{3}\left(x^2 + 4x - 1\right)^{3/2} + C$$

10. Let $u = x^5 + 2x$, so $\dfrac{du}{dx} = 5x^4 + 2$ and thus $du = \left(5x^4 + 2\right)dx$. Finding the integral:
    $$\int\left(5x^4 + 2\right)e^{x^5 + 2x}\,dx = \int e^u\,du = e^u + C = e^{x^5 + 2x} + C$$

11. Let $u = 2e^x + 3$, so $\dfrac{du}{dx} = 2e^x$ and thus $\dfrac{1}{2}du = e^x dx$. Finding the integral:
    $$\int \frac{e^x}{2e^x + 3}dx = \frac{1}{2}\int\frac{1}{u}du = \frac{1}{2}\ln|u| + C = \frac{1}{2}\ln\left(2e^x + 3\right) + C$$

12. Let $u = \ln x$, so $\dfrac{du}{dx} = \dfrac{1}{x}$ and thus $du = \dfrac{1}{x}dx$. Finding the integral:
    $$\int \frac{5\ln x}{x}dx = 5\int u\,du = \frac{5}{2}u^2 + C = \frac{5}{2}\left(\ln x\right)^2 + C$$

13. Let $u = x-1$, so $du = dx$ and $x = u+1$. Finding the integral:

$$\int \frac{x}{x-1}dx = \int \frac{u+1}{u}du = \int \left(1+\frac{1}{u}\right)du = u + \ln|u| + C = (x-1) + \ln|x-1| + C = x + \ln|x-1| + C$$

Note in the last step we absorbed the -1 constant into C.

14. Let $u = x-4$, so $du = dx$ and $x = u+4$. Finding the integral:

$$\int \frac{3x}{x-4}dx = 3\int \frac{u+4}{u}du = 3\int \left(1+\frac{4}{u}\right)du = 3u + 12\ln|u| + C = 3(x-4) + 12\ln|x-4| + C = 3x + 12\ln|x-4| + C$$

Note in the last step we absorbed the -12 constant into C.

15. Let $u = -0.13x$, so $\dfrac{du}{dx} = -0.13$ and thus $\dfrac{1}{-0.13}du = dx$. Finding the integral:

$$\int 52e^{-0.13x}dx = \frac{52}{-0.13}\int e^u du = -400e^u + C = -400e^{-0.13x} + C$$

16. Let $u = x^5$, so $\dfrac{du}{dx} = 5x^4$ and thus $\dfrac{1}{5}du = x^4 dx$. Finding the integral:

$$\int x^4 e^{x^5}dx = \frac{1}{5}\int e^u du = \frac{1}{5}e^u + C = \frac{1}{5}e^{x^5} + C$$

17. This integral represents the total number of people who have heard the news between days a and b.

18. Let $u = 3x-1$, so $\dfrac{du}{dx} = 3$ and thus $\dfrac{1}{3}du = dx$. When $x = 1$, $u = 2$, and when $x = 3$, $u = 8$. Finding the integral:

$$\int_1^3 (3x-1)^3 dx = \frac{1}{3}\int_2^8 u^3 du = \frac{1}{12}u^4 \Big|_2^8 = \frac{1}{12}(8^4 - 2^4) = 340$$

19. Let $u = 3x$, so $\dfrac{du}{dx} = 3$ and thus $\dfrac{1}{3}du = dx$. When $x = 0$, $u = 0$, and when $x = 1$, $u = 3$. Finding the integral:

$$\int_0^1 e^{3x}dx = \frac{1}{3}\int_0^3 e^u du = \frac{1}{3}e^u \Big|_0^3 = \frac{1}{3}(e^3 - 1)$$

20. Let $u = x^2 - 3$, so $\dfrac{du}{dx} = 2x$ and thus $du = 2xdx$. When $x = 0$, $u = -3$, and when $x = 1$, $u = -2$. Finding the integral:

$$\int_0^1 \frac{2x}{x^2-3}dx = \int_{-3}^{-2} \frac{1}{u}du = \ln|u|\Big|_{-3}^{-2} = \ln 2 - \ln 3 = \ln\frac{2}{3}$$

21. Let $u = 1-x^2$, so $\dfrac{du}{dx} = -2x$ and thus $-\dfrac{1}{2}du = xdx$. When $x = 0$, $u = 1$, and when $x = 1$, $u = 0$. Finding the integral:

$$\int_0^1 x\sqrt{1-x^2}dx = -\frac{1}{2}\int_1^0 u^{1/2}du = -\frac{1}{2}\cdot\frac{2}{3}u^{3/2}\Big|_1^0 = -\frac{1}{3}(0-1) = \frac{1}{3}$$

22. Let $u = x-3$, so $du = dx$ and $x = u+3$. When $x = 0$, $u = -3$, and when $x = 3$, $u = 0$. Finding the integral:

$$\int_0^3 x(x-3)^2 dx = \int_{-3}^0 (u+3)u^2 du = \int_{-3}^0 (u^3 + 3u^2)du = \left(\frac{1}{4}u^4 + u^3\right)\Big|_{-3}^0 = 0 - \left(\frac{81}{4} - 27\right) = \frac{27}{4}$$

23. Let $u = x^2$, so $\dfrac{du}{dx} = 2x$ and thus $\dfrac{1}{2}du = xdx$. When $x = 0$, $u = 0$, and when $x = 1$, $u = 1$. Finding the integral:

$$\int_0^1 xe^{x^2}dx = \frac{1}{2}\int_0^1 e^u du = \frac{1}{2}e^u \Big|_0^1 = \frac{1}{2}(e-1) \approx 0.86$$

24. Let $u = -x$, so $\dfrac{du}{dx} = -1$ and thus $-du = dx$. When $x = 0$, $u = 0$, and when $x = b$, $u = -b$.

Finding the integral: $\displaystyle\int_0^b e^{-x}dx = -\int_0^{-b} e^u du = -e^u\Big|_0^{-b} = -(e^{-b} - 1) = 1 - \frac{1}{e^b}$

Now taking limits: $\displaystyle\int_0^\infty e^{-x}dx = \lim_{b\to\infty}\left(1 - \frac{1}{e^b}\right) = 1$

---

25. Since there is a discontinuity at $x = 0$, we need to evaluate this as an improper integrals:

$$\int_0^1 \frac{1}{\sqrt[3]{x}} dx = \lim_{a \to 0^+} \int_a^1 \frac{1}{\sqrt[3]{x}} dx$$

Evaluating the integral:

$$\int_a^1 \frac{1}{\sqrt[3]{x}} dx = \int_a^1 x^{-1/3} dx = \frac{3}{2} x^{2/3} \Big|_a^1 = \frac{3}{2}\left(1 - a^{2/3}\right)$$

Now taking limits: $\int_0^1 \frac{1}{\sqrt[3]{x}} dx = \lim_{a \to 0^+} \int_a^1 \frac{1}{\sqrt[3]{x}} dx = \lim_{a \to 0^+} \left[\frac{3}{2}\left(1 - a^{2/3}\right)\right] = \frac{3}{2}$

26. Finding the integral:

$$\int_0^3 5{,}000\left(1 - e^{-0.04t}\right) dt = 5{,}000\int_0^3 dt - 5{,}000\int_0^3 e^{-0.04t} dt$$

$$= 5{,}000t - \frac{5{,}000}{-0.04} e^{-0.04t} \Big|_0^3$$

$$= 5{,}000t + 125{,}000 e^{-0.04t} \Big|_0^3$$

$$= \left(15{,}000 + 125{,}000 e^{-0.12}\right) - \left(0 + 125{,}000\right)$$

$$\approx 865.055$$

The company can expect to save about \$865,055 during the first three years.

Chapter 6
Applications of Integration

6.1 Area of Regions in the Plane

1.   Begin by sketching the region:

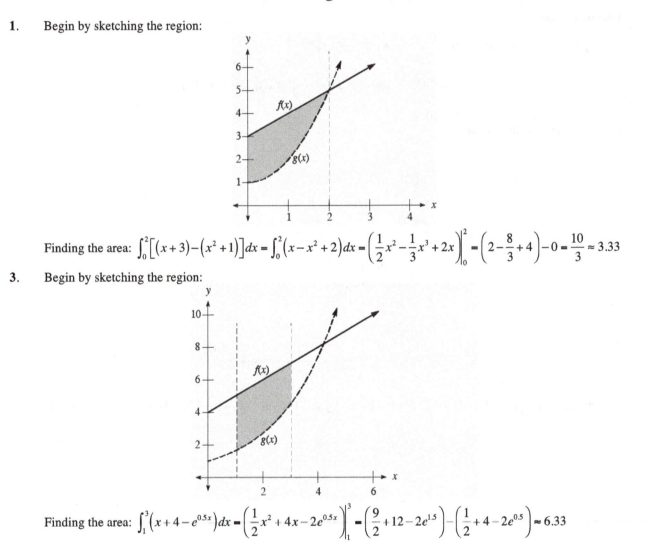

Finding the area: $\int_0^2 \left[(x+3)-(x^2+1)\right]dx = \int_0^2 (x-x^2+2)dx = \left(\frac{1}{2}x^2 - \frac{1}{3}x^3 + 2x\right)\Big|_0^2 = \left(2 - \frac{8}{3} + 4\right) - 0 = \frac{10}{3} \approx 3.33$

3.   Begin by sketching the region:

Finding the area: $\int_1^3 (x+4-e^{0.5x})dx = \left(\frac{1}{2}x^2 + 4x - 2e^{0.5x}\right)\Big|_1^3 = \left(\frac{9}{2} + 12 - 2e^{1.5}\right) - \left(\frac{1}{2} + 4 - 2e^{0.5}\right) \approx 6.33$

5. Begin by sketching the region:

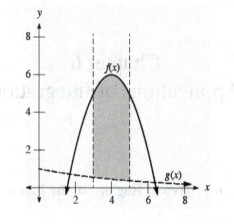

Finding the area:

$$\int_3^5 \left(-x^2 + 8x - 10 - e^{-0.2x}\right)dx = \left(-\frac{1}{3}x^3 + 4x^2 - 10x + 5e^{-0.2x}\right)\Bigg|_3^5$$

$$= \left(-\frac{125}{3} + 100 - 50 + 5e^{-1}\right) - \left(-9 + 36 - 30 + 5e^{-0.6}\right)$$

$$\approx 10.43$$

7. Begin by sketching the region:

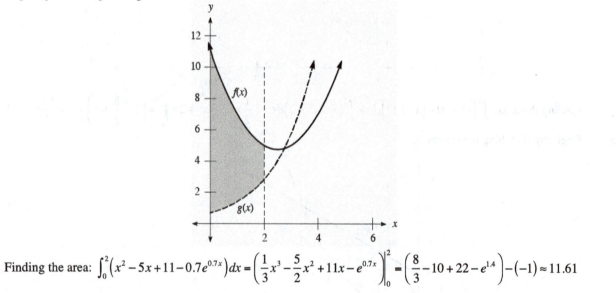

Finding the area: $\int_0^2 \left(x^2 - 5x + 11 - 0.7e^{0.7x}\right)dx = \left(\frac{1}{3}x^3 - \frac{5}{2}x^2 + 11x - e^{0.7x}\right)\Bigg|_0^2 = \left(\frac{8}{3} - 10 + 22 - e^{1.4}\right) - (-1) \approx 11.61$

9. Begin by sketching the region:

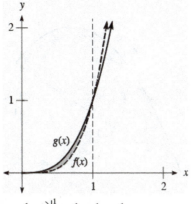

Finding the area: $\int_0^1 (x^3 - x^4)\,dx = \left(\dfrac{1}{4}x^4 - \dfrac{1}{5}x^5\right)\Big|_0^1 = \dfrac{1}{4} - \dfrac{1}{5} = \dfrac{1}{20} = 0.05$

11. Begin by sketching the region:

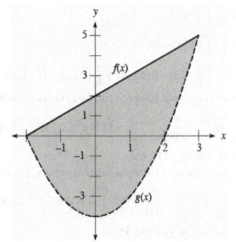

Finding the area:

$$\int_{-2}^3 \left[(x+2) - (x^2 - 4)\right]dx = \int_{-2}^3 (-x^2 + x + 6)\,dx$$

$$= \left(-\dfrac{1}{3}x^3 + \dfrac{1}{2}x^2 + 6x\right)\Big|_{-2}^3$$

$$= \left(-9 + \dfrac{9}{2} + 18\right) - \left(\dfrac{8}{3} + 2 - 12\right)$$

$$= \dfrac{125}{6}$$

$$\approx 20.83$$

13. Begin by sketching the region:

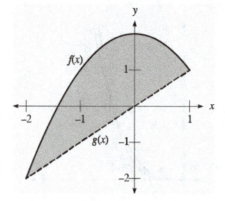

Finding the area:

$$\int_{-2}^{1}\left[\left(2-x^2\right)-\left(x\right)\right]dx = \int_{-2}^{1}\left(-x^2-x+2\right)dx = \left(-\frac{1}{3}x^3-\frac{1}{2}x^2+2x\right)\Big|_{-2}^{1} = \left(-\frac{1}{3}-\frac{1}{2}+2\right)-\left(\frac{8}{3}-2-4\right) = \frac{9}{2} = 4.5$$

15. The integral to represent this area is: $\int_c^d\left[g(x)-f(x)\right]dx$

17. The integral to represent this area is: $\int_d^c\left[k(x)-m(x)\right]dx$

19. The integral to represent this area is: $\int_a^b\left[f(x)-h(x)\right]dx + \int_b^c\left[h(x)-f(x)\right]dx$

21. The difference between the revenues is the area between the two curves. Finding the integral

$$\int_0^5\left(6e^{0.05t}-6e^{0.04t}\right)dt = \left(\frac{6}{0.05}e^{0.05t}-\frac{6}{0.04}e^{0.04t}\right)\Big|_0^5 = \left(120e^{0.25}-150e^{0.20}\right)-\left(120-150\right) \approx 0.872636$$

The difference in revenues is approximately $872,636.

23. The difference between the revenues is the area between the two curves. Finding the integral

$$\int_0^5\left(9e^{-0.02t}-9e^{-0.03t}\right)dt = \left(\frac{9}{-0.02}e^{-0.02t}-\frac{9}{-0.03}e^{-0.03t}\right)\Big|_0^5 = \left(-450e^{-0.10}+300e^{-0.15}\right)-\left(-450+300\right) \approx 1.035555$$

The difference in revenues is approximately $1,035,555.

25. a. Finding when the revenue equals the cost:

$$9,200-50t^2 = 2,000+150t^2$$
$$-200t^2 = -7,200$$
$$t^2 = 36$$
$$t = 6 \quad \left(\text{note } t > 0\right)$$

The company will keep the machine for 6 years before scrapping it.

b. Finding the total profit:

$$\int_0^6\left[\left(9,200-50t^2\right)-\left(2,000+150t^t\right)\right]dt = \int_0^6\left(7,200-200t^2\right)dt$$
$$= \left(7,200t-\frac{200}{3}t^3\right)\Big|_0^6$$
$$= \left(43,200-14,400\right)-0$$
$$= 28,800$$

The total profit will be $28,800 before the machine is scrapped.

---

27. Finding the difference in total projected revenue:

$$\int_0^5 \left(600{,}000e^{-0.012t} - 600{,}000e^{-0.018t}\right) dt$$

$$= \left(\frac{600{,}000e^{-0.012t}}{-0.012} - \frac{600{,}000e^{-0.018t}}{-0.018}\right)\Bigg|_0^5$$

$$= \left(-50{,}000{,}000e^{-0.012t} + \frac{100{,}000{,}000}{3}e^{-0.018t}\right)\Bigg|_0^5$$

$$= \left(-50{,}000{,}000e^{-0.06} + \frac{100{,}000{,}000}{3}e^{-0.09}\right) - \left(-50{,}000{,}000 + \frac{100{,}000{,}000}{3}\right)$$

$$\approx 42{,}813$$

The difference in projected revenues is approximately \$42,813.

29. Finding the difference between the projections:

$$\int_2^5 \left[\left(-x^4 + 72x^2\right) - \left(-x^4 + 80x^2\right)\right] dx = \int_2^5 -8x^2 \, dx$$

$$= -\frac{8}{3}x^3 \Bigg|_2^5$$

$$= -\frac{1000}{3} + \frac{64}{3}$$

$$= -312$$

The difference in the two projections is −\$312,000.

31. First find when the savings curve intersects the cost curve:

$$3.6t + 8 = 4.8t$$

$$8 = 1.2t$$

$$t = \frac{20}{3}$$

Now finding the area between the curves:

$$\int_0^{20/3} \left[(3.6t + 8) - (4.8t)\right] dt = \int_0^{20/3} (8 - 1.2t) \, dt$$

$$= \left(8t - 0.6t^2\right)\Bigg|_0^{20/3}$$

$$= \left(\frac{160}{3} - \frac{80}{3}\right) - 0$$

$$= \frac{80}{3}$$

$$\approx 26.667$$

The savings realized from purchasing new equipment is approximately \$26,667.

33. Finding the total change in store sales:

$$\int_0^{12} \left(500e^{0.12t} - 500e^{0.09t}\right) dt = \left(\frac{25{,}000}{6}e^{0.12t} - \frac{50{,}000}{9}e^{0.09t}\right)\Bigg|_0^{12}$$

$$= \left(\frac{25{,}000}{6}e^{1.44} - \frac{50{,}000}{9}e^{1.08}\right) - \left(\frac{25{,}000}{6} - \frac{50{,}000}{9}\right)$$

$$\approx 2{,}615.79$$

The sales will increase by approximately \$2,615.79 over the next 12 months.

35. Finding the area:

$$\int_0^e \left(200e^{0.12x} - 200e^{0.10x}\right) dx = \left(\frac{5{,}000}{3}e^{0.12x} - 2{,}000e^{0.10x}\right)\Bigg|_0^e = \left(\frac{5{,}000}{3}e^{0.12e} - 2{,}000e^{0.10e}\right) - \left(\frac{5{,}000}{3} - 2{,}000\right) \approx 18$$

The area is approximately 18 square units.

37. Finding the difference between the income flows:
$$\int_0^3 \left(450e^{0.08t} - 100e^{0.5t}\right)dt = \left(5,625e^{0.08t} - 200e^{0.5t}\right)\Big|_0^3 = \left(5,625e^{0.24} - 200e^{1.5}\right) - \left(5,625 - 200\right) \approx 829.44$$

The difference in total income is approximately \$829.44.

39. Solving the equation:
$$250 = 1,450 - 3x^2$$
$$3x^2 = 1,200$$
$$x^2 = 400$$
$$x = \pm 20$$

41. Solving the equation:
$$0.004x^2 = 25 - 0.005x^2$$
$$0.009x^2 = 25$$
$$x^2 = \frac{25,000}{9}$$
$$x = \pm 52.70$$

43. Finding the integral: $\int_0^{52.70} \left(11.11 - 0.004x^2\right)dx = \left(11.11x - \frac{0.004}{3}x^3\right)\Big|_0^{52.70} \approx 585.497 - 195.151 \approx 390.35$

45. Finding the integral:
$$\int_0^{200} \left[114 - \left(0.0002x^2 + 0.03x + 100\right)\right]dx = \int_0^{200}\left(14 - 0.0002x^2 - 0.03x\right)dx$$
$$= \left(14x - \frac{0.0002}{3}x^3 - 0.015x^2\right)\Big|_0^{200}$$
$$= 2,800 - \frac{1,600}{3} - 600 - 0$$
$$= \frac{5,000}{3}$$
$$\approx 1,666.67$$

6.2 Consumer's and Producer's Surplus

1. Finding the consumer's surplus:
$$\int_0^{15}\left[\left(750 - 2x^2\right) - 300\right]dx = \int_0^{15}\left(450 - 2x^2\right)dx = \left(450x - \frac{2}{3}x^3\right)\Big|_0^{15} = 6,750 - 2,250 = \$4,500$$

3. Finding the producer's surplus:
$$\int_0^{18}\left[434 - \left(x^2 + 5x + 20\right)\right]dx = \int_0^{18}\left(414 - 5x - x^2\right)dx = \left(414x - \frac{5}{2}x^2 - \frac{1}{3}x^3\right)\Big|_0^{18} = 7,452 - 810 - 1,944 = \$4,698$$

5. Finding the consumer's surplus:
$$\int_0^{38/3}\left[\left(50 - 2x\right) - \frac{74}{3}\right]dx = \int_0^{38/3}\left(\frac{76}{3} - 2x\right)dx = \left(\frac{76}{3}x - x^2\right)\Big|_0^{38/3} = \frac{2,888}{9} - \frac{1,444}{9} = \frac{1,444}{9} \approx \$160.44$$

Finding the producer's surplus:
$$\int_0^{38/3}\left[\frac{74}{3} - \left(12 + x\right)\right]dx = \int_0^{38/3}\left(\frac{38}{3} - x\right)dx = \left(\frac{38}{3}x - \frac{1}{2}x^2\right)\Big|_0^{38/3} = \frac{1,444}{9} - \frac{1,444}{18} = \frac{1,444}{18} \approx \$80.22$$

7. The consumer's surplus is \$581,250 when the price is \$240.
9. The producer's surplus is \$45.85 when the price is \$1,226.

11. First we need to find the number of units when the price is $2,500:
$$10,000 - 3x^2 = 2,500$$
$$-3x^2 = -7,500$$
$$x^2 = 2,500$$
$$x = 50$$

Finding the consumer's surplus:
$$\int_0^{50} \left[\left(10,000 - 3x^2 \right) - 2,500 \right] dx = \int_0^{50} \left(7,500 - 3x^2 \right) dx = \left(7,500x - x^3 \right) \Big|_0^{50} = 375,000 - 125,000 = \$250,000$$

13. First we need to find the number of units when the price is $5:
$$-0.01x^2 - 0.2x + 8 = 5$$
$$-0.01x^2 - 0.2x + 3 = 0$$
$$x^2 + 20x - 300 = 0$$
$$(x + 30)(x - 10) = 0$$
$$x = 10 \quad (x = -30 \text{ is impossible})$$

Finding the consumer's surplus:
$$\int_0^{10} \left[\left(-0.01x^2 - 0.2x + 8 \right) - 5 \right] dx = \int_0^{10} \left(-0.01x^2 - 0.2x + 3 \right) dx$$
$$= \left(\frac{-0.01x^3}{3} - 0.1x^2 + 3x \right) \Big|_0^{10}$$
$$= -\frac{10}{3} - 10 + 30$$
$$= \frac{50}{3}$$
$$\approx \$16.67$$

15. First we need to find the number of units when the price is $20:
$$\frac{400}{x + 5} = 20$$
$$400 = 20(x + 5)$$
$$20 = x + 5$$
$$x = 15$$

Finding the consumer's surplus:
$$\int_0^{15} \left(\frac{400}{x + 5} - 20 \right) dx = \left(400 \ln |x + 5| - 20x \right) \Big|_0^{15} = 400 \ln 20 - 300 - 400 \ln 5 \approx \$254.52$$

17. First we need to find the number of units when the price is $15:
$$\sqrt{285 - 2.4x} = 15$$
$$285 - 2.4x = 225$$
$$-2.4x = -60$$
$$x = 25$$

Finding the consumer's surplus:
$$\int_0^{25} \left(\sqrt{285 - 2.4x} - 15 \right) dx = \left(-\frac{5}{18} (285 - 2.4x)^{3/2} - 15x \right) \Big|_0^{25} = -\frac{5}{18} (225)^{3/2} - 375 + \frac{5}{18} (285)^{3/2} \approx \$23.99$$

$$\left(285 - 2.4x \right)^{1/2}$$
$$= \frac{1}{2} (285 - 2.4x)^{3/2}$$

19. First we need to find the number of units when the price is $298:

$$150 + 200e^{-0.001x} = 298$$
$$200e^{-0.001x} = 148$$
$$e^{-0.001x} = 0.74$$
$$-0.001x = \ln 0.74$$
$$x = \frac{\ln 0.74}{-0.001} \approx 301.105$$

Finding the consumer's surplus:

$$\int_0^{301.105} \left(150 + 200e^{-0.001x} - 298\right)dx = \int_0^{301.105}\left(200e^{-0.001x} - 148\right)dx$$
$$= \left(-200{,}000e^{-0.001x} - 148x\right)\Big|_0^{301.105}$$
$$\approx -200{,}000e^{-0.301105} - 44{,}563.54 + 200{,}000$$
$$\approx \$7{,}436.45$$

21. First we need to find the number of units when the price is $47:

$$11 + 0.36x^2 = 47$$
$$0.36x^2 = 36$$
$$x^2 = 100$$
$$x = 10$$

Finding the producer's surplus:

$$\int_0^{10}\left[47 - \left(11 + 0.36x^2\right)\right]dx = \int_0^{10}\left(36 - 0.36x^2\right)dx = \left(36x - 0.12x^3\right)\Big|_0^{10} = 360 - 120 = \$240$$

23. First find the price when $x = 1{,}500$: $S(1{,}500) = 0.8e^{0.002(1{,}500)} = 0.8e^3$

Finding the producer's surplus:

$$\int_0^{1{,}500}\left(0.8e^3 - 0.8e^{0.002x}\right)dx = \left(0.8e^3 x - 400e^{0.002x}\right)\Big|_0^{1{,}500} = 1{,}200e^3 - 400e^3 + 400 \approx \$16{,}468.43$$

25. First find the equilibrium point by setting the supply equal to the demand:

$$0.03x^2 - 1.6x + 136 = -0.01x^2 + 220$$
$$0.04x^2 - 1.6x - 84 = 0$$
$$x^2 - 40x - 2{,}100 = 0$$
$$(x + 30)(x - 70) = 0$$
$$x = 70 \quad (x \neq -30)$$

Finding the equilibrium price: $D(70) = -0.01(70)^2 + 220 = \$171$

Finding the consumer's surplus:

$$\int_0^{70}\left(-0.01x^2 + 220 - 171\right)dx = \int_0^{70}\left(-0.01x^2 + 49\right)dx$$
$$= \left(-\frac{1}{300}x^3 + 49x\right)\Big|_0^{70}$$
$$= -\frac{3{,}430}{3} + 3{,}430$$
$$\approx \$2{,}286.67$$

Finding the producer's surplus:

$$\int_0^{70}\left[171-\left(0.03x^2-1.6x+136\right)\right]dx = \int_0^{70}\left(-0.03x^2+1.6x+35\right)dx$$

$$= \left(-0.01x^3+0.8x^2+35x\right)\Big|_0^{70}$$

$$= -3,430+3,920+2,450$$

$$= \$2,940$$

27. Finding the producer's surplus:

$$\int_0^{200}\left[114-\left(0.0002x^2+0.03x+100\right)\right]dx = \int_0^{200}\left(-0.0002x^2-0.03x+14\right)dx$$

$$= \left(\frac{-0.0002x^3}{3}-0.015x^2+14x\right)\Big|_0^{200}$$

$$= -533.33-600+2,800$$

$$\approx \$1,666.67$$

29. Setting the supply equal to the price:

$$200+40e^{0.04x}=495.56$$

$$40e^{0.04x}=295.56$$

$$e^{0.04x}=7.389$$

$$0.04x=\ln 7.389$$

$$x=\frac{\ln 7.389}{0.04}\approx 50$$

The equilibrium point is $(50, 495.56)$.

31. Evaluating the integral: $\int_0^6 100e^{0.005t}\, dt = 20,000e^{0.005t}\Big|_0^6 = 20,000e^{0.03}-20,000 \approx 609.09$

33. Computing the value: $Ae^{-rt}=15,000e^{-0.08(10)}=15,000e^{-0.8}\approx 6,739.93$

35. Evaluating the integral: $\int_0^5 10,000e^{-0.07t}\, dt = \dfrac{10,000e^{-0.07t}}{-0.07}\Big|_0^5 = -\dfrac{10,000}{0.07}e^{-0.35}+\dfrac{10,000}{0.07}\approx 42,187.42$

37. First find the integral: $\int_0^T 70,000e^{-0.065t}\, dt = \dfrac{70,000e^{-0.065t}}{-0.065}\Big|_0^T = -\dfrac{70,000}{0.065}e^{-0.065T}+\dfrac{70,000}{0.065}$

Now taking limits: $\int_0^\infty 70,000e^{-0.065t}\, dt = \lim_{T\to\infty}\left(-\dfrac{70,000}{0.065}e^{-0.065T}+\dfrac{70,000}{0.065}\right)=\dfrac{70,000}{0.065}\approx 1,076,923.08$

6.3 Annuities and Money Streams

1. Finding the amount accumulated: $A = \int_0^3 200{,}000\, dt = 200{,}000t\big|_0^3 = \$600{,}000$

3. Finding the amount accumulated: $A = 90{,}000e^{0.10(6)} = 90{,}000e^{0.6} \approx \$163{,}990.69$

5. First convert the rate to a monthly rate: $r = \dfrac{0.06}{12} = 0.005$

   Finding the future value: $A = \int_0^{24} 300e^{0.005t}\, dt = 60{,}000e^{0.005t}\big|_0^{24} = 60{,}000e^{0.12} - 60{,}000 \approx \$7{,}649.81$

7. Finding the present value: $P = 120{,}000e^{-0.10(5)} = 120{,}000e^{-0.5} \approx \$72{,}783.68$

9. Finding the future value: $A = \int_0^{10} 20{,}000e^{0.12t}\, dt = \dfrac{20{,}000}{0.12}e^{0.12t}\Big|_0^{10} = \dfrac{20{,}000}{0.12}e^{1.2} - \dfrac{20{,}000}{0.12} \approx \$386{,}686.15$

11. Finding the present value: $P = \int_0^{10} 20{,}000e^{-0.12t}\, dt = -\dfrac{20{,}000}{0.12}e^{-0.12t}\Big|_0^{10} = -\dfrac{20{,}000}{0.12}e^{-1.2} + \dfrac{20{,}000}{0.12} \approx \$116{,}467.63$

13. First find the integral: $\int_0^T 1{,}000e^{-0.05t}\, dt = -20{,}000e^{-0.05t}\big|_0^T = -20{,}000e^{-0.05T} + 20{,}000$

    Finding the present value: $P = \lim_{T\to\infty} \int_0^T 1{,}000e^{-0.05t}\, dt = \lim_{T\to\infty}\left(-20{,}000e^{-0.05T} + 20{,}000\right) = \$20{,}000$

15. They should invest the present value of $362{,}854.49$ because it is cheaper than the cost of the machine, which is $375{,}000$.

17. If the machine costs less than the present value of $151{,}189.37$, they should buy the machine.

19. The present value of $2{,}914.18$ indicates the amount needed, without further investments, to reach the future value after 10 months. The future value of $3{,}089.23$ indicates the future value after the 10 months of investments.

21. The present value of $11{,}868.61$ indicates the amount needed, without further investments, to reach the future value after 15 years. The future value of $29{,}192.06$ indicates the future value after the 15 years of investments.

23. Finding the amount accumulated: $A = \int_0^3 300{,}000\, dt = 300{,}000t\big|_0^3 = \$900{,}000$

25. Finding the amount accumulated: $A = 69{,}880.70e^{0.06(20)} = 69{,}880.70e^{1.2} \approx \$232{,}012$

27. First convert the rate to a monthly rate: $r = \dfrac{0.06}{12} = 0.005$

    Finding the future value: $A = \int_0^{240} 500e^{0.005t}\, dt = 100{,}000e^{0.005t}\big|_0^{240} = 100{,}000e^{1.2} - 100{,}000 \approx \$232{,}012$

29. Finding the present value: $P = 5{,}000{,}000e^{-0.05(3)} = 5{,}000{,}000e^{-0.15} \approx \$4{,}303{,}540$

31. a. Finding the future value:

    $$e^{0.05(3)}\int_0^3 200{,}000e^{-0.05t}\, dt = e^{0.15}\cdot(-4{,}000{,}000)e^{-0.05t}\big|_0^3$$
    $$= e^{0.15}\left(-4{,}000{,}000e^{-0.15} + 4{,}000{,}000\right)$$
    $$= 4{,}000{,}000e^{0.15} - 4{,}000{,}000$$
    $$\approx \$647{,}337$$

    The income stream will be worth $647{,}337$ at the end of 3 years.

    b. Finding the present value: $\int_0^3 200{,}000e^{-0.05t}\, dt = -4{,}000{,}000e^{-0.05t}\big|_0^3 = -4{,}000{,}000e^{-0.15} + 4{,}000{,}000 \approx \$557{,}168$

    If $557{,}168$ were invested today at 5% continuous interest for 3 years, its value would reach $647{,}337$.

33. **a.** Finding the future value:

$$e^{0.10(6)}\int_0^6 1{,}000{,}000e^{-0.10t}\,dt = e^{0.6}\cdot(-10{,}000{,}000)e^{-0.10t}\Big|_0^6$$

$$= e^{0.6}\left(-10{,}000{,}000e^{-0.6}+10{,}000{,}000\right)$$

$$= 10{,}000{,}000e^{0.6}-10{,}000{,}000$$

$$\approx \$8{,}221{,}188$$

The income stream will be worth $8,221,188 at the end of 6 years.

b. Finding the present value:

$$\int_0^6 1{,}000{,}000e^{-0.10t}\,dt = -10{,}000{,}000e^{-0.10t}\Big|_0^6 = -10{,}000{,}000e^{-0.6}+10{,}000{,}000 \approx \$4{,}511{,}884$$

If $4,511,884 were invested today at 10% continuous interest for 6 years, its value would reach $8,221,188.

35. **a.** The machine will produce 3($100,000) = $300,000 of direct income (with no investment of income stream).

b. Finding the future value:

$$e^{0.04(3)}\int_0^3 100{,}000e^{-0.04t}\,dt = e^{0.12}\cdot(-2{,}500{,}000)e^{-0.04t}\Big|_0^3$$

$$= e^{0.12}\left(-2{,}500{,}000e^{-0.12}+2{,}500{,}000\right)$$

$$= 2{,}500{,}000e^{0.12}-2{,}500{,}000$$

$$\approx \$318{,}742$$

The machine will produce $318,742 of income if the income stream is continuously reinvested.

c. Finding the present value:

$$\int_0^3 100{,}000e^{-0.04t}\,dt = -2{,}500{,}000e^{-0.04t}\Big|_0^3 = -2{,}500{,}000e^{-0.12}+2{,}500{,}000 \approx \$282{,}699$$

If $282,699 were invested today at 4% continuous interest for 3 years, its value would reach $318,742.

d. The company should buy the machine, since its cost ($250,000) is less than the present value ($282,699).

37. **a.** Finding the amount accumulated: $A = \int_0^{60} 10{,}000\,dt = 10{,}000t\Big|_0^{60} = \$600{,}000$

b. First convert the rate to a monthly rate: $r = \dfrac{0.04}{12} = \dfrac{0.01}{3}$

Finding the future value: $A = \int_0^{60} 10{,}000e^{\left(\frac{0.01}{3}t\right)}\,dt = 3{,}000{,}000e^{\left(\frac{0.01}{3}t\right)}\Big|_0^{60} = 3{,}000{,}000e^{0.2}-3{,}000{,}000 \approx \$664{,}208$

c. The difference is $64,208.

39. First find the integral: $\int_0^T 70{,}000e^{-0.07t}\,dt = -1{,}000{,}000e^{-0.07t}\Big|_0^T = -1{,}000{,}000e^{-0.07T}+1{,}000{,}000$

Finding the present value: $P = \lim_{T\to\infty}\int_0^T 70{,}000e^{-0.07t}\,dt = \lim_{T\to\infty}\left(-1{,}000{,}000e^{-0.07T}+1{,}000{,}000\right) = \$1{,}000{,}000$

An investment of $1,000,000 would have to be made now in order to match the total income from the lease.

41. **a.** Finding the future value:

$$e^{0.085(5)}\int_0^5\left(36{,}450e^{-0.052t}\right)e^{-0.085t}\,dt = e^{0.425}\int_0^5 36{,}450e^{-0.137t}\,dt$$

$$= e^{0.425}\cdot\left(-\frac{36{,}450}{0.137}\right)e^{-0.137t}\Big|_0^5$$

$$= e^{0.425}\left(-\frac{36{,}450}{0.137}e^{-0.685}+\frac{36{,}450}{0.137}\right)$$

$$= \frac{36{,}450}{0.137}e^{0.425}-\frac{36{,}450}{0.137}e^{-0.26}$$

$$\approx \$201{,}816$$

b. Finding the present value:

$$\int_0^5 \left(36{,}450e^{-0.052t}\right)e^{-0.085t}\,dt = \int_0^5 36{,}450e^{-0.137t}\,dt = -\frac{36{,}450}{0.137}e^{-0.137t}\Big|_0^5 = -\frac{36{,}450}{0.137}e^{-0.685} + \frac{36{,}450}{0.137} \approx \$131{,}941$$

43. First find the integral: $\displaystyle\int_0^T 62{,}575e^{-0.085t}\,dt = -\frac{62{,}575}{0.085}e^{-0.085t}\Big|_0^T = -\frac{62{,}575}{0.085}e^{-0.07T} + \frac{62{,}575}{0.085}$

Finding the present value: $\displaystyle P = \lim_{T\to\infty}\int_0^T 62{,}575e^{-0.085t}\,dt = \lim_{T\to\infty}\left(-\frac{62{,}575}{0.085}e^{-0.085T} + \frac{62{,}575}{0.085}\right) \approx \$736{,}176$

An investment of \$736,176 would have to be made now in order to match the total income from the lease.

45. Let $u = 1 + x^2$, so $du = 2x\,dx$ and $\dfrac{1}{2}du = x\,dx$. Evaluating the integral:

$$\int \frac{x}{1+x^2}\,dx = \frac{1}{2}\int\frac{1}{u}\,du = \frac{1}{2}\ln|u| + C = \frac{1}{2}\ln\left(x^2+1\right) + C$$

47. Solving for C:

$$\frac{-1}{-0.1} = 2^3 + C$$
$$10 = 8 + C$$
$$C = 2$$

49. Solving for C:

$$3^4 = 10(-4)^2 + C$$
$$81 = 160 + C$$
$$C = -79$$

Now solving for y:

$$y^4 = 10x^2 - 79$$
$$y = \pm\sqrt[4]{10x^2 - 79}$$

51. Solving for C:

$$200(100)^{1/2} = 0 + C$$
$$C = 2{,}000$$

Now solving for p:

$$200p^{1/2} = t + 2{,}000$$
$$p^{1/2} = \frac{t + 2{,}000}{200}$$
$$p = \left(\frac{t + 2{,}000}{200}\right)^2$$

6.4 Differential Equations

1. Solving the differential equation:

$$\frac{dy}{dx} = x^4$$

$$dy = x^4 dx$$

$$\int dy = \int x^4 \, dx$$

$$y = \frac{1}{5}x^5 + C$$

3. Solving the differential equation:

$$\frac{dy}{dx} = 0$$

$$dy = 0 \, dx$$

$$\int dy = \int 0 \, dx$$

$$y = C$$

5. Solving the differential equation:

$$y\frac{dy}{dx} = 2x$$

$$y \, dy = 2x \, dx$$

$$\int y \, dy = \int 2x \, dx$$

$$\frac{1}{2}y^2 = x^2 + C$$

$$y^2 = 2x^2 + C$$

$$y = \pm\sqrt{2x^2 + C}$$

7. Solving the differential equation:

$$\frac{dy}{dx} = xe^{x^2}$$

$$dy = xe^{x^2} dx$$

$$\int dy = \int xe^{x^2} \, dx$$

$$y = \frac{1}{2}e^{x^2} + C$$

9. Solving the differential equation:

$$(1+x)\frac{dy}{dx} = x$$

$$dy = \frac{x}{x+1} dx$$

$$\int dy = \int \frac{x}{x+1} dx$$

$$y = x - \ln|x+1| + C$$

11. Solving the differential equation:

$$x\frac{dy}{dx} = y\ln y$$

$$\frac{dy}{y\ln y} = \frac{1}{x} dx$$

$$\int \frac{dy}{y\ln y} = \int \frac{1}{x} dx$$

$$\ln|\ln y| = \ln|x| + C$$

$$\ln y = Cx$$

$$y = e^{Cx}$$

13. Solving the differential equation:

$$\frac{dy}{dx} = \frac{x^2}{y^2}$$

$$y^2 dy = x^2 dx$$

$$\int y^2 \, dy = \int x^2 \, dx$$

$$\frac{1}{3}y^3 = \frac{1}{3}x^3 + C$$

$$y^3 = x^3 + C$$

Now evaluating to find C:

$$y^3 = x^3 + C$$

$$4^3 = 0^3 + C$$

$$C = 64$$

The solution is:

$$y^3 = x^3 + 64$$

$$y = \sqrt[3]{x^3 + 64}$$

15. Solving the differential equation:

$$\frac{dy}{dx} = \frac{1}{xy}$$

$$y\,dy = \frac{1}{x}\,dx$$

$$\int y\,dy = \int \frac{1}{x}\,dx$$

$$\frac{1}{2}y^2 = \ln|x| + C$$

$$y^2 = 2\ln|x| + C$$

Now evaluating to find C:

$$y^2 = 2\ln|x| + C$$

$$4^2 = 2\ln 1 + C$$

$$16 = 0 + C$$

$$C = 16$$

The solution is:

$$y^2 = 2\ln|x| + 16$$

$$y = \sqrt{2\ln|x| + 16}$$

Note that we chose the positive root here since y is positive.

17. Solving the differential equation:

$$\frac{dy}{dx} = xe^x - x$$

$$dy = \left(xe^x - x\right)dx$$

$$\int dy = \int \left(xe^x - x\right)dx$$

$$y = \int xe^x\,dx - \int x\,dx$$

Making the substitutions:

$$u = x \qquad\qquad dv = e^x dx$$

$$du = dx \qquad\qquad v = e^x$$

Integrating by parts:

$$y = \int xe^x\,dx - \int x\,dx = xe^x - \int e^x\,dx - \int x\,dx = xe^x - e^x - \frac{1}{2}x^2 + C = (x-1)e^x - \frac{1}{2}x^2 + C$$

Now evaluating to find C:

$$y = (x-1)e^x - \frac{1}{2}x^2 + C$$

$$-1 = (0-1)e^0 - \frac{1}{2}(0)^2 + C$$

$$-1 = -1 + C$$

$$C = 0$$

The solution is: $y = (x-1)e^x - \frac{1}{2}x^2$

19. Solving the differential equation:

$$2xy\frac{dy}{dx} = 1 + y^2$$

$$\frac{y}{y^2+1}dy = \frac{1}{2x}dx$$

$$\int \frac{y}{y^2+1}dy = \int \frac{1}{2x}dx$$

$$\frac{1}{2}\ln(y^2+1) = \frac{1}{2}\ln|x| + C$$

$$\ln(y^2+1) = \ln|x| + C$$

$$y^2 = Cx - 1$$

Now evaluating to find C:

$$y^2 = Cx - 1$$

$$3^2 = 5C - 1$$

$$10 = 5C$$

$$C = 2$$

The solution is:

$$y^2 = 2x - 1$$

$$y = \sqrt{2x-1}$$

21. Solving the differential equation:

$$\frac{dP}{dt} = kP$$

$$\frac{1}{P}dP = kdt$$

$$\int \frac{1}{P}dP = \int k\,dt$$

$$\ln|P| = kt + C$$

$$P = Ce^{kt}$$

Now evaluating to find C:

$$P = Ce^{kt}$$

$$P_0 = Ce^0$$

$$C = P_0$$

The solution is: $P = P_0 e^{kt}$

23. Finding the derivative: $\dfrac{dy}{dx} = 4x$

25. Finding the derivative: $\dfrac{dy}{dx} = Ce^{x^2} \cdot 2x = 2x\left(Ce^{x^2}\right) = 2xy$

27. Finding the derivative: $\dfrac{dy}{dx} = Ce^x = y$

29. Finding the derivative implicitly:

$$\frac{1}{x} - \frac{1}{y^2} \cdot \frac{dy}{dx} = 0$$

$$-\frac{1}{y^2} \cdot \frac{dy}{dx} = -\frac{1}{x}$$

$$\frac{dy}{dx} = \frac{y^2}{x}$$

31. This is a separable differential equation.

33. This is not a separable differential equation.

35. This is a separable differential equation.

37. This is not a separable differential equation.

39. Solving the differential equation:

$$\frac{dQ}{dt} = k(160 - Q)$$

$$\frac{dQ}{160 - Q} = k\,dt$$

$$\int \frac{dQ}{160 - Q} = \int k\,dt$$

$$-\ln|160 - Q| = kt + C$$

$$160 - Q = Ce^{-kt}$$

$$Q = 160 - Ce^{-kt}$$

Since $Q = 0$ when $t = 0$:

$$Q = 160 - Ce^{-kt}$$

$$0 = 160 - Ce^{0}$$

$$C = 160$$

So $Q = 160 - 160e^{-kt}$. Since $Q = 35$ when $t = 4$:

$$Q = 160 - 160e^{-kt}$$

$$35 = 160 - 160e^{-4k}$$

$$-125 = -160e^{-4k}$$

$$e^{-4k} = \frac{25}{32}$$

$$-4k = \ln\left(\frac{25}{32}\right)$$

$$k = -\frac{1}{4}\ln\left(\frac{25}{32}\right) \approx 0.061715$$

So $Q = 160 - 160e^{-0.061715t}$. Now substituting $t = 18$: $Q = 160 - 160e^{-0.061715(18)} \approx 107$

The average student can type 107 words per minute after 18 weeks of instruction.

41. Solving the differential equation ($t = 0$ corresponds to 2009):

$$\frac{dN}{dt} = 0.04N$$

$$\frac{dN}{N} = 0.04dt$$

$$\int \frac{dN}{N} = \int 0.04\, dt$$

$$\ln|N| = 0.04t + C$$

$$N = Ce^{0.04t}$$

Since $N = 135$ when $t = 0$:

$$N = Ce^{0.04t}$$

$$135 = Ce^{0}$$

$$C = 135$$

So $N = 135e^{0.04t}$. Now substituting $t = 5$: $N = 135e^{0.04(5)} \approx 165$

In 2014 the county can expect to have 165 white-collar crimes reported.

43. Solving the differential equation:

$$\frac{dP}{dh} = kP$$

$$\frac{dP}{P} = kdh$$

$$\int \frac{dP}{P} = \int k\, dh$$

$$\ln|P| = kh + C$$

$$P = Ce^{kh}$$

Since $P = P_0$ when $h = 0$:

$$P = Ce^{kh}$$

$$P_0 = Ce^{0}$$

$$C = P_0$$

So $P = P_0 e^{kh}$. Since $P = \frac{1}{2}P_0$ when $h = 18{,}000$:

$$P = P_0 e^{kh}$$

$$\frac{1}{2}P_0 = P_0 e^{18{,}000k}$$

$$e^{18{,}000k} = \frac{1}{2}$$

$$18{,}000k = \ln\frac{1}{2}$$

$$k = \frac{1}{18{,}000}\ln\frac{1}{2}$$

So $P = P_0 e^{\left(\frac{1}{18{,}000}\ln\frac{1}{2}\right)h}$. Now substituting $h = 10{,}000$: $P = P_0 e^{\left(\frac{1}{18{,}000}\ln\frac{1}{2}\right)10{,}000} \approx 0.68P_0$

At 10,000 feet the pressure is 68% of the pressure at sea level.

45. Solving the differential equation:

$$\frac{dC}{dx} = kC$$

$$\frac{dC}{C} = k\,dx$$

$$\int \frac{dC}{C} = \int k\,dx$$

$$\ln|C| = kx + A$$

$$C = Ae^{kx}$$

Since $C = 4$ when $x = 0$:

$$C = Ae^{kx}$$

$$4 = Ae^0$$

$$A = 4$$

So $C = 4e^{kx}$. Since $C = 10$ when $x = 50$:

$$C = 4e^{kx}$$

$$10 = 4e^{50k}$$

$$e^{50k} = 2.5$$

$$50k = \ln 2.5$$

$$k = \frac{1}{50}\ln 2.5 \approx 0.018326$$

So $C = 4e^{0.018326x}$. Now substituting $x = 70$: $C = 4e^{0.018326(70)} \approx 14.4$

With the addition of 20 new homes, the pollutants entering the river would be 14.4 ppm.

47. Solving the differential equation:

$$\frac{dP}{dt} = kP^{1/3}$$

$$P^{-1/3}\,dP = k\,dt$$

$$\int P^{-1/3}\,dP = \int k\,dt$$

$$\frac{3}{2}P^{2/3} = kt + C$$

$$P^{2/3} = \frac{2}{3}(kt + C)$$

$$P = (kt + C)^{3/2}$$

Note that the constants were changed in the last step.
Since $P = 60$ when $t = 0$:

$$P = (kt + C)^{3/2}$$

$$60 = C^{3/2}$$

$$C = 60^{2/3} \approx 15.326$$

So $P = (kt + 15.326)^{3/2}$. Since $P = 53$ when $t = 4$:

$$P = (kt + 15.326)^{3/2}$$
$$53 = (4k + 15.326)^{3/2}$$
$$53^{2/3} = 4k + 15.326$$
$$4k = 53^{2/3} - 15.326$$
$$k = \frac{1}{4}(53^{2/3} - 15.326) \approx -0.3041$$

So $P = (-0.3041t + 15.326)^{3/2}$. Substituting $t = 16$: $P = (-0.3041 \cdot 16 + 15.326)^{3/2} \approx 34\%$

Approximately 34% of the registered voters can be expected to vote in 2016.

49. Solving the differential equation:

$$\frac{da}{dt} = 0.2a^{3/4}$$
$$a^{-3/4} da = 0.2 dt$$
$$\int a^{-3/4} da = \int 0.2 dt$$
$$4a^{1/4} = 0.2t + C$$
$$a^{1/4} = 0.05t + C$$
$$a = (0.05t + C)^4$$

Since $a = 20$ when $t = 0$:

$$a = (0.05t + C)^4$$
$$20 = C^4$$
$$C = 20^{1/4} \approx 2.11$$

So $a = (0.05t + 2.11)^4$.

51. Solving the differential equation:

$$\frac{dy}{dx} = -\frac{y}{x^2}$$
$$\frac{1}{y} dy = -x^{-2} dx$$
$$\int \frac{1}{y} dy = -\int x^{-2} dx$$
$$\ln|y| = \frac{1}{x} + C$$
$$y = Ce^{1/x}$$

53. Solving the differential equation:

$$\frac{dy}{dx} = x^2 e^x - x$$

$$dy = \left(x^2 e^x - x\right) dx$$

$$\int dy = \int \left(x^2 e^x - x\right) dx$$

$$y = \int x^2 e^x \, dx - \int x \, dx$$

Making the substitutions:

$$u = x^2 \qquad\qquad dv = e^x dx$$

$$du = 2x dx \qquad\qquad v = e^x$$

Integrating by parts: $y = \int x^2 e^x \, dx - \int x \, dx + C = x^2 e^x - \int 2x e^x \, dx - \frac{1}{2} x^2 + C$

Making the substitutions:

$$u = x \qquad\qquad dv = e^x dx$$

$$du = x dx \qquad\qquad v = e^x$$

Integrating by parts: $y = x^2 e^x - 2\left(x e^x - \int e^x \, dx\right) - \frac{1}{2} x^2 + C = x^2 e^x - 2x e^x + 2e^x - \frac{1}{2} x^2 + C = \left(x^2 - 2x + 2\right) e^x - \frac{1}{2} x^2 + C$

Now evaluating to find C:

$$y = \left(x^2 - 2x + 2\right) e^x - \frac{1}{2} x^2 + C$$

$$2 = 2e^0 - \frac{1}{2}(0)^2 + C$$

$$2 = 2 + C$$

$$C = 0$$

The solution is: $y = \left(x^2 - 2x + 2\right) e^x - \frac{1}{2} x^2$

55. Solving the differential equation:

$$\frac{dP}{dt} = kP$$

$$\frac{dP}{P} = k dt$$

$$\int \frac{dP}{P} = \int k \, dt$$

$$\ln|P| = kt + C$$

$$P = C e^{dt}$$

57. Evaluating the integral: $\int \frac{dP}{L-P} = -\ln|L-P| + C$

59. Solving for k:

$$160 = 30 + 190 e^{-k(20)}$$

$$130 = 190 e^{-20k}$$

$$e^{-20k} = \frac{13}{19}$$

$$-20k = \ln\left(\frac{13}{10}\right)$$

$$k = -\frac{1}{20} \ln\left(\frac{13}{19}\right) \approx 0.019$$

61. Solving for k:

$$4{,}500 = \frac{114{,}000{,}000}{3{,}800 + 26{,}200 e^{-30{,}000 k (24)}}$$

$$45 = \frac{11{,}400}{38 + 262 e^{-720{,}000 k}}$$

$$1{,}710 + 11{,}790 e^{-720{,}000 k} = 11{,}400$$

$$11{,}790 e^{-720{,}000 k} = 9{,}690$$

$$e^{-720{,}000 k} = \frac{969}{1{,}179}$$

$$-720{,}000 k = \ln\left(\frac{969}{1{,}179}\right)$$

$$k = -\frac{1}{720{,}000}\ln\left(\frac{969}{1{,}179}\right) \approx 0.000000272$$

6.5 Applications of Differential Equations

1. Evaluating to find C:

$$P = Ce^{kt}$$

$$P_0 = Ce^{0}$$

$$C = P_0$$

The solution is: $P = P_0 e^{kt}$

3. Finding the value of k:

$$P = 4{,}000 - 3{,}920 e^{-kt}$$

$$2{,}800 = 4{,}000 - 3{,}920 e^{-k(2)}$$

$$-1{,}200 = -3{,}920 e^{-2k}$$

$$e^{-2k} = \frac{15}{49}$$

$$-2k = \ln\left(\frac{15}{49}\right)$$

$$k = -\frac{1}{2}\ln\left(\frac{15}{49}\right) \approx 0.592$$

5. This is unlimited growth, since the rate is proportional to the current population with no upper limit.

7. This is unlimited growth, since the rate is proportional to the current population with no upper limit.

9. This is logistic growth, since the rate is proportional to both the infected individuals and the non-infected individuals.

11. This is logistic growth, since the rate is proportional to both the people who know the procedure and those who need to learn the procedure.

13.   **a.**   This is the limited growth model with $L = 68$ and $P_0 = 40$, so the equation is

$P = 68 + (40 - 68)e^{-kt} = 68 - 28e^{-kt}$. Substituting $P = 45$ when $t = 15$:

$$45 = 68 - 28e^{-k(15)}$$

$$-23 = -28e^{-15k}$$

$$e^{-15k} = \frac{23}{28}$$

$$-15k = \ln\left(\frac{23}{28}\right)$$

$$k = -\frac{1}{15}\ln\left(\frac{23}{28}\right) \approx 0.01311$$

So the function is $P = 68 - 28e^{-0.01311t}$.

     b.   Substituting $t = 30$: $P = 68 - 28e^{-0.01311(30)} \approx 49Y$

15.   **a.**   This is the logistic growth model with $L = 650$ and $P_0 = 45$, so the equation is

$P = \dfrac{(45)(650)}{45 + (650 - 45)e^{-650kt}} = \dfrac{29{,}250}{45 + 605e^{-650kt}}$. Substituting $P = 160$ when $t = 2$:

$$160 = \frac{29{,}250}{45 + 605e^{-650k(2)}}$$

$$160\left(45 + 605e^{-1{,}300k}\right) = 29{,}250$$

$$45 + 605e^{-1{,}300k} = 182.8125$$

$$605e^{-1{,}300k} = 137.8125$$

$$e^{-1{,}300k} = \frac{137.8125}{605}$$

$$-1{,}300k = \ln\left(\frac{137.8125}{605}\right)$$

$$k = -\frac{1}{1{,}300}\ln\left(\frac{137.8125}{605}\right) \approx 0.00113795$$

So the function is $P = \dfrac{29{,}250}{45 + 605e^{-650(0.00113795)t}} = \dfrac{29{,}250}{45 + 605e^{-0.739667t}}$.

     b.   Substituting $t = 4$: $P = \dfrac{29{,}250}{45 + 605e^{-0.739667(4)}} \approx 383$ units

17. **a.** This is the logistic growth model with $L = 2,500$ and $P_0 = 1$, so the equation is

$$P = \frac{(1)(2,500)}{1+(2,500-1)e^{-2,500kt}} = \frac{2,500}{1+2,499e^{-2,500kt}}$$. Substituting $P = 40$ when $t = 2$:

$$40 = \frac{2,500}{1+2,499e^{-2,500k(2)}}$$

$$40\left(1+2,499e^{-5,000k}\right) = 2,500$$

$$1+2,499e^{-5,000k} = 62.5$$

$$2,499e^{-5,000k} = 61.5$$

$$e^{-5,000k} = \frac{61.5}{2,499}$$

$$-5,000k = \ln\left(\frac{61.5}{2,499}\right)$$

$$k = -\frac{1}{5,000}\ln\left(\frac{61.5}{2,499}\right) \approx 0.0007409$$

So the function is $P = \dfrac{2,500}{1+2,499e^{-2,500(0.0007409)t}} = \dfrac{2,500}{1+2,499e^{-1.852304t}}$.

b. Substituting $t = 6$: $P = \dfrac{2,500}{1+2,499e^{-1.852304(6)}} \approx 2,410$ people

19. This is the unlimited growth model, so the equation is $P = P_0 e^{kt}$.

Substituting $P = \dfrac{1}{2}P_0$ when $t = 5,770$:

$$\frac{1}{2}P_0 = P_0 e^{k(5,770)}$$

$$e^{5,770k} = 0.5$$

$$5,770k = \ln 0.5$$

$$k = \frac{\ln 0.5}{5,770} \approx -0.00012013$$

So the function is $P = P_0 e^{-0.00012013t}$.

Substituting $P = \dfrac{1}{5}P_0$:

$$\frac{1}{5}P_0 = P_0 e^{-0.00012013t}$$

$$0.2 = e^{-0.00012013t}$$

$$-0.00012013t = \ln 0.2$$

$$t = \frac{\ln 0.2}{-0.00012013} \approx 13,398 \text{ years old}$$

21. This is the logistic growth model with $L = 1$ and $P_0 = 0.06$, so the equation is

$$P = \frac{(0.06)(1)}{0.06 + (1 - 0.06)e^{-1kt}} = \frac{0.06}{0.06 + 0.94e^{-kt}}$$. Substituting $P = 0.60$ when $t = 4$:

$$0.60 = \frac{0.06}{0.06 + 0.94e^{-k(4)}}$$

$$0.6\left(0.06 + 0.94e^{-4k}\right) = 0.06$$

$$0.06 + 0.94e^{-4k} = 0.1$$

$$0.94e^{-4k} = 0.04$$

$$e^{-4k} = \frac{2}{27}$$

$$-4k = \ln\left(\frac{2}{27}\right)$$

$$k = -\frac{1}{4}\ln\left(\frac{2}{27}\right) \approx 0.78925$$

So the function is $P = \dfrac{0.06}{0.06 + 0.94e^{-0.78925t}}$.

Substituting $P = 0.99$:

$$0.99 = \frac{0.06}{0.06 + 0.94e^{-0.78925t}}$$

$$0.99\left(0.06 + 0.94e^{-0.78925t}\right) = 0.06$$

$$0.06 + 0.94e^{-0.78925t} = \frac{2}{33}$$

$$0.94e^{-0.78925t} = \frac{2}{33} - 0.06$$

$$e^{-0.78925t} \approx 0.00064475$$

$$-0.78925t = \ln 0.00064475$$

$$t = \frac{\ln 0.00064475}{-0.78925} \approx 9 \text{ hours}$$

23. Converting the two solutions:

$$P = e^{-kt}\left(P_0 + L\left(e^{kt} - 1\right)\right)$$

$$P = e^{-kt}\left(P_0 + Le^{kt} - L\right)$$

$$P = P_0 e^{-kt} + L - Le^{-kt}$$

$$P = L + \left(P_0 - L\right)e^{-kt}$$

25. Solving and converting the two solutions:

$$P = \frac{P_0 L e^{kLt}}{P_0\left(e^{kLt}-1\right)+L}$$

$$P = \frac{P_0 L e^{kLt}}{P_0 e^{kLt} - P_0 + L}$$

$$P = \frac{P_0 L}{\left(P_0 e^{kLt} - P_0 + L\right)e^{-kLt}}$$

$$P = \frac{P_0 L}{P_0 - P_0 e^{-kLt} + L e^{-kLt}}$$

$$P = \frac{P_0 L}{P_0 + \left(L - P_0\right)e^{-kLt}}$$

27. Finding the sum: $0.6957 + 0.2618 + 0.0394 + 0.0030 + 0.0001 + 0.000002 = 1.000002$.

29. Evaluating the integral: $\dfrac{3}{124}\displaystyle\int_1^3 x^2\,dx = \dfrac{1}{124}x^3\Big|_1^3 = \dfrac{27}{124} - \dfrac{1}{124} = \dfrac{26}{124} = \dfrac{13}{62} \approx 0.2097$

31. Evaluating the integral: $\displaystyle\int_{120}^{150} 0.01e^{-0.01x}\,dx = -e^{-0.01x}\Big|_{120}^{150} = -e^{-1.5} + e^{-1.2} \approx 0.0781$

6.6 Probability

1. Both conditions fail. Since $f(1) = \dfrac{1-2}{16} = -\dfrac{1}{16} < 0$, condition 1 fails. Finding the integral:

$$\int_{-\infty}^{\infty} f(x)\,dx = \frac{1}{16}\int_1^5 (x-2)\,dx = \frac{1}{16}\left(\frac{1}{2}x^2 - 2x\right)\Big|_1^5 = \frac{1}{16}\left(\frac{25}{2}-10\right) - \frac{1}{16}\left(\frac{1}{2}-2\right) = \frac{5}{32} + \frac{3}{32} = \frac{1}{4} \neq 1$$

So condition 2 fails also.

3. Both conditions fail. Since $f(2) = 4(2)^2 = 16 > 1$, condition 1 fails. Finding the integral:

$$\int_{-\infty}^{\infty} f(x)\,dx = \int_{-2}^{2} 4x^2\,dx = \frac{4}{3}x^3\Big|_{-2}^{2} = \frac{32}{3} + \frac{32}{3} = \frac{64}{3} \neq 1$$

So condition 2 fails also.

5. Both conditions fail. Since $f(-12) = \dfrac{5}{12}(-12) = -5 < 0$, condition 1 fails. Finding the integral:

$$\int_{-\infty}^{\infty} f(x)\,dx = \int_{-12}^{12} \frac{5}{12}x\,dx = \frac{5}{24}x^2\Big|_{-12}^{12} = 30 - 30 = 0 \neq 1$$

So condition 2 fails also.

7. Finding the integral: $\displaystyle\int_{-\infty}^{\infty} f(x)\,dx = \int_1^8 kx^{1/3}\,dx = \frac{3}{4}kx^{4/3}\Big|_1^8 = \frac{3}{4}k(16-1) = \frac{45}{4}k$

Setting this expression equal to 1:

$$\frac{45}{4}k = 1$$

$$k = \frac{4}{45}$$

9. Finding the integral: $\int_{-\infty}^{\infty} f(x)dx = \int_1^5 kx\,dx = \frac{1}{2}kx^2\Big|_1^5 = \frac{1}{2}k(25-1) = 12k$

Setting this expression equal to 1:
$$12k = 1$$
$$k = \frac{1}{12}$$

11. Finding the integral:
$$\int_{-\infty}^{\infty} f(x)dx = \int_4^9 kx^{-1/2}\,dx = 2kx^{1/2}\Big|_4^9 = 2k(3-2) = 2k$$

Setting this expression equal to 1:
$$2k = 1$$
$$k = \frac{1}{2}$$

13. Finding the probability: $\int_0^1 \frac{xe^{-x^2/2}}{2}\,dx = -\frac{1}{2}e^{-x^2/2}\Big|_0^1 = -\frac{1}{2}\left(e^{-1/2}-1\right) = \frac{1}{2}\left(1-e^{-1/2}\right) \approx 0.1967$

15. Finding the probability: $\int_1^{10} \frac{1}{9x}\,dx = \frac{1}{9}\ln x\Big|_1^{10} = \frac{1}{9}(\ln 10 - 0) = \frac{1}{9}\ln 10 \approx 0.2558$

17. Finding the probability: $\int_5^{10} 0.04e^{-0.04x}\,dx = -e^{-0.04x}\Big|_5^{10} = -\left(e^{-0.4}-e^{-0.2}\right) = e^{-0.2} - e^{-0.4} \approx 0.1484$

19. Finding the probability: $\int_2^4 \frac{3}{40}\left(x^2-2x\right)dx = \frac{3}{40}\left(\frac{1}{3}x^3 - x^2\right)\Big|_2^4 = \frac{3}{40}\left(\frac{64}{3}-16\right) - \frac{3}{40}\left(\frac{8}{3}-4\right) = \frac{3}{40}\left(\frac{16}{3}+\frac{4}{3}\right) = \frac{1}{2}$

21. Approximately 952 out of 10,000 air travelers are expected to be exposed to between 5.00 and 5.50 mrem of radiation while flying across the continental United States.

23. Between 2.4 and 4.4 inches of rain will occur in February in a particular county 571 out of 1,000 times.

25. Approximately 151 out of 1,000 phone calls in a certain city will last longer than 16 minutes.

27. Approximately 6 out of 1,000 people will react to a stimulus after more than 0.2 seconds.

29. Finding the integral: $\int_{-\infty}^{\infty} p(x)dx = \int_2^5 \frac{3}{100}x^2\,dx = \frac{1}{100}x^3\Big|_2^5 = \frac{125}{100} - \frac{8}{100} = \frac{117}{100} \neq 1$

So condition 2 fails.

31. The probability density function is $f(x) = \frac{1}{150-0} = \frac{1}{150}$.

   a. Finding the probability: $\int_0^{20} \frac{1}{150}\,dx = \frac{1}{150}x\Big|_0^{20} = \frac{20}{150} - \frac{0}{150} = \frac{2}{15} \approx 0.1333$

   b. Finding the probability: $\int_{20}^{150} \frac{1}{150}\,dx = \frac{1}{150}x\Big|_{20}^{150} = \frac{150}{150} - \frac{20}{150} = \frac{13}{15} \approx 0.8667$

   c. Finding the probability: $\int_0^{75} \frac{1}{150}\,dx = \frac{1}{150}x\Big|_0^{75} = \frac{75}{150} - \frac{0}{150} = \frac{1}{2} = 0.5000$

   d. Finding the probability: $\int_{75}^{150} \frac{1}{150}\,dx = \frac{1}{150}x\Big|_{75}^{150} = \frac{150}{150} - \frac{75}{150} = \frac{1}{2} = 0.5000$

33. The probability density function is $f(x) = \frac{1}{3,600-0} = \frac{1}{3,600}$.

   Since 12 minutes = 720 seconds, finding the probability: $\int_0^{720} \frac{1}{3,600}\,dx = \frac{1}{3,600}x\Big|_0^{720} = \frac{720}{3,600} - \frac{0}{3,600} = \frac{1}{5} = 0.2$

   Out of 1,000 requests, we would expect: $0.2(1,000) = 200$ requests.

35. The probability density function is $f(x) = \dfrac{1}{72-24} = \dfrac{1}{48}$.

Finding the probability: $\displaystyle\int_{66}^{72} \dfrac{1}{48}\,dx = \dfrac{1}{48}x\Big|_{66}^{72} = \dfrac{72}{48} - \dfrac{66}{48} = \dfrac{1}{8} = 0.125$

Approximately 12.5% of people can be expected to completely digest and eliminate a meal in the final 6 hours.

37. Finding the probability: $\displaystyle\int_{0}^{100} 0.0002e^{-0.0002x}\,dx = -e^{-0.0002x}\Big|_{0}^{100} = -e^{-0.02} + 1 \approx 0.0198$

Approximately 1.98% of the devices will last less than 100 hours.

39. a. Finding the probability: $\displaystyle\int_{0}^{1} \dfrac{1}{5}e^{-\frac{1}{5}x}\,dx = -e^{-\frac{1}{5}x}\Big|_{0}^{1} = -e^{-\frac{1}{5}} + 1 \approx 0.1813$

Approximately 18.13% of the devices will last less than 1 year.

b. Finding the probability: $\displaystyle\int_{1}^{\infty} \dfrac{1}{5}e^{-\frac{1}{5}x}\,dx = \lim_{t\to\infty}\int_{1}^{t} \dfrac{1}{5}e^{-\frac{1}{5}x}\,dx = \lim_{t\to\infty}\left(-e^{-\frac{1}{5}x}\Big|_{1}^{t}\right) = \lim_{t\to\infty}\left(-e^{-\frac{1}{5}t} + e^{-\frac{1}{5}}\right) = e^{-\frac{1}{5}} \approx 0.8187$

Approximately 81.87% of the devices will last more than 1 year.

41. a. First find the z-value: $z = \dfrac{10-20}{4} = -\dfrac{10}{4} = -2.5$

Using the normal probability table: $P(x < 10) = P(z < -2.5) = 0.5 - 0.4938 = 0.0062$

b. First find the z-values:

$z = \dfrac{20-20}{4} = \dfrac{0}{4} = 0$ $\qquad\qquad z = \dfrac{35-20}{4} = \dfrac{15}{4} = 3.75$

Using the normal probability table: $P(20 < x < 35) = P(0 < z < 3.75) = 0.4999 - 0 = 0.4999$

43. First find the z-values:

$z = \dfrac{190-180}{50} = \dfrac{10}{50} = 0.20$ $\qquad\qquad z = \dfrac{200-180}{50} = \dfrac{20}{50} = 0.40$

Using the normal probability table: $P(190 < x < 200) = P(0.20 < z < 0.40) = 0.1554 - 0.0793 = 0.0761$

45. First find the z-values:

$z = \dfrac{17.65-19.29}{1} = -\dfrac{1.64}{1} = -1.64$ $\qquad\qquad z = \dfrac{20.93-19.29}{1} = \dfrac{1.64}{1} = 1.64$

Using the normal probability table: $P(17.65 < x < 20.93) = P(-1.64 < z < 1.64) = 0.4495 + 0.4495 = 0.8990$

We would expect 89.9% of U.S. women to fit comfortably in the seats.

47. a. Finding the probability: $\displaystyle\int_{10}^{15} \dfrac{16.535}{16(t+1)^2}\,dt = -\dfrac{16.535}{16(t+1)}\Big|_{10}^{15} = -\dfrac{16.535}{256} + \dfrac{16.535}{176} \approx 0.0294$

b. Finding the probability: $\displaystyle\int_{0}^{5} \dfrac{16.535}{16(t+1)^2}\,dt = -\dfrac{16.535}{16(t+1)}\Big|_{0}^{5} = -\dfrac{16.535}{96} + \dfrac{16.535}{16} \approx 0.8612$

c. Finding the probability: $\displaystyle\int_{30}^{60} \dfrac{16.535}{16(t+1)^2}\,dt = -\dfrac{16.535}{16(t+1)}\Big|_{30}^{60} = -\dfrac{16.535}{976} + \dfrac{16.535}{496} \approx 0.0164$

49. a. Finding the probability: $\displaystyle\int_{0}^{3} 0.024e^{-0.024t}\,dt = -e^{-0.024t}\Big|_{0}^{3} = -e^{-0.072} + 1 \approx 0.0695$

b. Finding the probability:

$\displaystyle\int_{10}^{\infty} 0.024e^{-0.024t}\,dt = \lim_{t\to\infty}\int_{10}^{t} 0.024e^{-0.024t}\,dt = \lim_{t\to\infty}\left(-e^{-0.024t}\Big|_{10}^{t}\right) = \lim_{t\to\infty}\left(-e^{-0.024t} + e^{-0.24}\right) = e^{-0.24} \approx 0.7866$

51. Note the normal probability table produces values which are slightly different from calculator or Wolfram|Alpha values.

    a. First find the z-value: $z = \dfrac{5-12}{3.3} = -\dfrac{7}{3.3} \approx -2.12$

       Using the normal probability table: $P(x < 5) = P(z < -2.12) = 0.5 - 0.4830 = 0.0170$

    b. First find the z-values:

$$z = \frac{8-12}{3.3} = -\frac{4}{3.3} \approx -1.21 \qquad\qquad z = \frac{12-12}{3.3} = \frac{0}{3.3} = 0$$

       Using the normal probability table: $P(8 < x < 12) = P(-1.21 < z < 0) = 0.3869$

    c. First find the z-value: $z = \dfrac{20-12}{3.3} = \dfrac{8}{3.3} \approx 2.42$

       Using the normal probability table: $P(x > 20) = P(z > 2.42) = 0.5 - 0.4922 = 0.0078$

53. Evaluating the expression: $2,300(80) + 1,400(16) + 900(20) - 3(80)^2 - (16)^2 - (20)^2 = 204,544$

55. Evaluating the expression: $4(85)^{3/5}(20)^{2/5} \approx 190.60$

57. Substituting the expressions: $4(kx)^{3/5}(ky)^{2/5} = 4k^{3/5}x^{3/5} \cdot k^{2/5}y^{2/5} = k \cdot 4x^{3/5}y^{2/5}$

Chapter 6 Test

1. Begin by sketching the region:

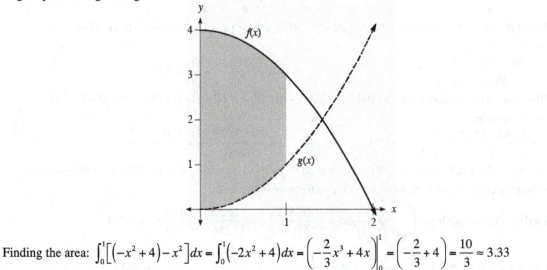

Finding the area: $\displaystyle\int_0^1 \left[(-x^2 + 4) - x^2 \right] dx = \int_0^1 (-2x^2 + 4)\, dx = \left(-\frac{2}{3}x^3 + 4x \right)\Big|_0^1 = \left(-\frac{2}{3} + 4 \right) = \frac{10}{3} \approx 3.33$

2. First find where the two curves intersect:
$$3x^2 + 3x + 5 = 4x^2 + x - 3$$
$$-x^2 + 2x + 8 = 0$$
$$x^2 - 2x - 8 = 0$$
$$(x+2)(x-4) = 0$$
$$x = -2, 4$$

Now finding the area:
$$\int_{-2}^{4}\left[(3x^2+3x+5)-(4x^2+x-3)\right]dx = \int_{-2}^{4}(-x^2+2x+8)dx$$
$$= \left(-\frac{1}{3}x^3 + x^2 + 8x\right)\Big|_{-2}^{4}$$
$$= \left(-\frac{64}{3} + 16 + 32\right) - \left(\frac{8}{3} + 4 - 16\right)$$
$$= \frac{80}{3} + \frac{28}{3}$$
$$= 36$$

3. First find the equilibrium point by setting the supply equal to the demand:
$$0.25x^2 = 300 - 0.5x^2$$
$$0.75x^2 = 300$$
$$x^2 = 400$$
$$x = 20 \quad (x \neq -20)$$

Finding the equilibrium price: $D(20) = 300 - 0.5(20)^2 = \$100$

The equilibrium point is (20, 100).

4. Finding the consumer's surplus using the equilibrium point (20, 100):
$$\int_{0}^{20}(300 - 0.5x^2 - 100)dx = \int_{0}^{20}(200 - 0.5x^2)dx$$
$$= \left(200x - \frac{0.5}{3}x^3\right)\Big|_{0}^{20}$$
$$= 4,000 - \frac{4,000}{3}$$
$$\approx \$2,666.67$$

The total savings to the consumer is approximately \$2,666.67.

5. Finding the producer's surplus: $\int_{0}^{20}(100 - 0.25x^2)dx = \left(100x - \frac{0.25}{3}x^3\right)\Big|_{0}^{20} = 2,000 - \frac{2,000}{3} \approx \$1,333.33$

The total savings to the producer is approximately \$1,333.33

6. Finding the amount accumulated: $A = \int_{0}^{3}100,000\,dt = 100,000t\Big|_{0}^{3} = \$300,000$

7. Finding the amount accumulated: $A = 5,000e^{0.06(10)} = 5,000e^{0.6} \approx \$9,110.59$

8. First convert the rate to a monthly rate: $r = \dfrac{0.05}{12}$

Finding the future value: $A = \int_{0}^{360}1,000e^{\left(\frac{0.05}{12}\right)t}\,dt = 240,000e^{\left(\frac{0.05}{12}\right)t}\Big|_{0}^{360} = 240,000e^{1.5} - 240,000 \approx \$835,605$

9. Finding the present value: $P = 100,000e^{-0.05(18)} = 100,000e^{-0.9} \approx \$40,656.97$

10. Finding the future value: $A = \int_0^{10} 50,000e^{0.04t}\,dt = 1,250,000e^{0.04t}\Big|_0^{10} = 1,250,000e^{0.4} - 1,250,000 \approx \$614,780.87$

11. Finding the present value: $\int_0^{10} 100,000e^{-0.05t}\,dt = -2,000,000e^{-0.05t}\Big|_0^{10} = -2,000,000e^{-0.5} + 2,000,000 \approx \$786,939$

12. First find the integral: $\int_0^T 20,000e^{-0.05t}\,dt = -400,000e^{-0.05t}\Big|_0^T = -400,000e^{-0.05T} + 400,000$

   Finding the present value: $P = \lim_{T \to \infty} \int_0^T 20,000e^{-0.05t}\,dt = \lim_{T \to \infty}\left(-400,000e^{-0.05T} + 400,000\right) = \$400,000$

13. Finding the derivative implicitly:

$$2x - 2y\frac{dy}{dx} = 0$$

$$-2y\frac{dy}{dx} = -2x$$

$$\frac{dy}{dx} = \frac{x}{y}$$

   No, the equation is not a solution to the differential equation.

14. Separating variables:

$$\frac{dy}{dx} = \frac{y}{x}\sqrt{1-x^2}$$

$$x\,dy = y\sqrt{1-x^2}\,dx$$

$$\frac{1}{y}\,dy = \frac{\sqrt{1-x^2}}{x}\,dx$$

$$\frac{dy}{dx} = \frac{x}{y}$$

15. Solving the differential equation:

$$\frac{dy}{dx} = \frac{2x}{y^2}$$

$$y^2\,dy = 2x\,dx$$

$$\int y^2\,dy = \int 2x\,dx\,dx$$

$$\frac{1}{3}y^3 = x^2 + C$$

$$y^3 = 3x^2 + C$$

   Now evaluating to find C:

$$y^3 = 3x^2 + C$$

$$4^3 = 3(1)^2 + C$$

$$64 = 3 + C$$

$$61 = C$$

   The solution is:

$$y^3 = 3x^2 + 61$$

$$y = \left(3x^2 + 61\right)^{1/3}$$

16. Solving the differential equation:

$$\frac{dy}{dx} = y^2\left(1+e^x\right)$$

$$y^{-2}dy = \left(1+e^x\right)dx$$

$$\int y^{-2}dy = \int\left(1+e^x\right)dx$$

$$-\frac{1}{y} = x + e^x + C$$

$$y = -\frac{1}{x + e^x + C}$$

Now evaluating to find C:

$$y = -\frac{1}{x + e^x + C}$$

$$1 = -\frac{1}{0 + e^0 + C}$$

$$1 = -\frac{1}{C+1}$$

$$C + 1 = -1$$

$$C = -2$$

The solution is:

$$y = -\frac{1}{x + e^x - 2}$$

$$y = \frac{1}{2 - x - e^x}$$

17. This is the unlimited growth model with $P_0 = 500,000$, so the equation is $P = 500,000e^{kt}$.

Substituting $P = 1,000,000$ when $t = 10$:

$$1,000,000 = 500,000e^{k(10)}$$

$$e^{10k} = 2$$

$$10k = \ln 2$$

$$k = \frac{\ln 2}{10} \approx 0.069315$$

So the function is $P = 500,000e^{0.069315t}$.

Substituting $P = 2,000,000$:

$$2,000,000 = 500,000e^{0.069315t}$$

$$e^{0.069315t} = 4$$

$$e^{0.069315t} = \ln 4$$

$$t = \frac{\ln 4}{0.069315} = 20$$

The population will grow to 2,000,000 in 20 hours.

18. This is the limited growth model with $L = 120$ and $P_0 = 50$, so the equation is $P = 120 + (50 - 120)e^{-kt} = 120 - 70e^{-kt}$.
Substituting $P = 60$ when $t = 200$:

$$60 = 120 - 70e^{-k(200)}$$

$$-60 = -70e^{-200k}$$

$$e^{-200k} = \frac{60}{70}$$

$$-200k = \ln\left(\frac{60}{70}\right)$$

$$k = -\frac{1}{200}\ln\left(\frac{60}{70}\right) \approx 0.0007708$$

So the function is $P = 120 - 70e^{-0.0007708t}$.
Substituting $P = 70$:

$$70 = 120 - 70e^{-0.0007708t}$$

$$-50 = -70e^{-0.0007708t}$$

$$e^{-0.0007708t} = \frac{5}{7}$$

$$-0.0007708t = \ln\left(\frac{5}{7}\right)$$

$$t = \frac{\ln\left(\frac{5}{7}\right)}{-0.0007708} \approx 437$$

The temperature will reach $70°$ in approximately 437 hours.

19. This is the logistic growth model with $L = 5{,}000{,}000$ and $P_0 = 10{,}000$, so the equation is

$$P = \frac{(10{,}000)(5{,}000{,}000)}{10{,}000 + (5{,}000{,}000 - 10{,}000)e^{-5{,}000{,}000kt}} = \frac{50{,}000{,}000{,}000}{10{,}000 + 4{,}990{,}000e^{-5{,}000{,}000kt}} = \frac{5{,}000{,}000}{1 + 499e^{-5{,}000{,}000kt}}.$$

Substituting $P = 80{,}800$ when $t = 2$:

$$80{,}800 = \frac{5{,}000{,}000}{1 + 499e^{-5{,}000{,}000k(2)}}$$

$$80{,}800\left(1 + 499e^{-10{,}000{,}000k}\right) = 5{,}000{,}000$$

$$1 + 499e^{-10{,}000{,}000k} = \frac{6{,}250}{101}$$

$$499e^{-10{,}000{,}000k} = \frac{6{,}149}{101}$$

$$e^{-10{,}000{,}000k} = \frac{6{,}149}{50{,}399}$$

$$-10{,}000{,}000k = \ln\left(\frac{6{,}149}{50{,}399}\right)$$

$$k = -\frac{1}{10{,}000{,}000}\ln\left(\frac{6{,}149}{50{,}399}\right) \approx 0.00000021$$

So the function is $P = \dfrac{5{,}000{,}000}{1 + 499e^{-5{,}000{,}000(0.00000021)t}} = \dfrac{5{,}000{,}000}{1 + 499e^{-1.0518409t}}$.

Substituting $P = 4,000,000$:

$$4,000,000 = \frac{5,000,000}{1 + 499e^{-1.0518409t}}$$

$$4,000,000\left(1 + 499e^{-1.0518409t}\right) = 5,000,000$$

$$1 + 499e^{-1.0518409t} = 1.25$$

$$499e^{-1.0518409t} = 0.25$$

$$e^{-1.0518409t} = \frac{1}{1,996}$$

$$-1.0518409t = \ln\left(\frac{1}{1,996}\right)$$

$$t = \frac{\ln\left(\dfrac{1}{1,996}\right)}{-1.0518409} \approx 7.2 \text{ days}$$

It will take approximately 7.2 days for the bacterial population to reach 4,000,000.

20. Approximately 1,855 out of 10,000 males aged 15-20 will buy a particular style of advertised clothes.

21. Finding the integral: $\displaystyle\int_{-\infty}^{\infty} f(x)\,dx = \int_{5}^{10} k\sqrt{2}x^{-2}\,dx = -\left.\frac{k\sqrt{2}}{x}\right|_{5}^{10} = -k\sqrt{2}\left(\frac{1}{10} - \frac{1}{5}\right) = \frac{k\sqrt{2}}{10}$

Setting this expression equal to 1:

$$\frac{k\sqrt{2}}{10} = 1$$

$$k = \frac{10}{\sqrt{2}}$$

$$k = \frac{10\sqrt{2}}{2} = 5\sqrt{2}$$

22. Finding the probability: $\displaystyle\int_{0}^{5} \frac{4}{789}\left(15 - x^{1/3}\right)dx = \left.\frac{4}{789}\left(15x - \frac{3}{4}x^{4/3}\right)\right|_{0}^{5} = \frac{4}{789}\left(75 - \frac{3}{4}(5)^{4/3}\right) \approx 0.3477$

They will need to replace approximately 35% of the batteries.

23. The probability density function is $f(x) = \dfrac{1}{35}$.

Finding the probability: $\displaystyle\int_{0}^{15} \frac{1}{35}\,dx = \left.\frac{1}{35}x\right|_{0}^{15} = \frac{15}{35} - \frac{0}{35} = \frac{3}{7} \approx 0.4286$

24. Finding the probability: $\displaystyle\int_{5}^{\infty} 0.36e^{-0.36x}\,dx = \lim_{t\to\infty}\int_{5}^{t} 0.36e^{-0.36x}\,dx = \lim_{t\to\infty}\left(-e^{-0.36x}\Big|_{5}^{t}\right) = \lim_{t\to\infty}\left(-e^{-0.36t} + e^{-1.8}\right) = e^{-1.8} \approx 0.1653$

25. First find the z-value: $z = \dfrac{6,000,000 - 5,000,000}{1,000,000} = \dfrac{1,000,000}{1,000,000} = 1.00$

Using the normal probability table: $P(x < 6,000,000) = P(z < 1) = 0.5 + 0.3413 = 0.8413$

Chapter 7
Calculus of Functions of Several Variables

7.1 Functions of Several Variables

1. Evaluating the function: $f(1,4) = 2(1)^3 - (4)^2 = 2 - 16 = -14$

3. Evaluating the function: $f(200,60) = 4(200)^{1/3}(60)^{2/3} \approx 358.51$

5. Evaluating the function: $P(20,27) = \sqrt{3(20)^3 - 2(27)^2} = \sqrt{22,542} \approx 150.14$

7. Evaluating the function: $f(5,8) = e^{0.01(3\cdot5-5\cdot8)} \ln(5\cdot8 - 3\cdot5) = e^{-0.25} \ln 25 \approx 2.51$

9. Evaluating the function: $P(12,2,20) = e^{\sqrt{12+3\cdot20}} - e^{\sqrt{2+\ln 2}} = e^{\sqrt{72}} - e^{\sqrt{2+\ln 2}} \approx 4,837.80$

11. Evaluating the function: $f(0,1) = 1^{e^0 + \ln 1} = 1^1 = 1$

13. Simplifying the expression:

$$\frac{f(x+h,y) - f(x,y)}{h} = \frac{\left[(x+h)^2 - y^2\right] - \left[x^2 - y^2\right]}{h} = \frac{x^2 + 2xh + h^2 - y^2 - x^2 + y^2}{h} = \frac{2xh + h^2}{h} = 2x + h$$

15. Finding the limit:

$$\lim_{h\to0} \frac{f(x,y+h) - f(x,y)}{h} = \lim_{h\to0} \frac{\left[(y+h)^2 + 4x(y+h)\right] - \left[y^2 + 4xy\right]}{h}$$

$$= \lim_{h\to0} \frac{y^2 + 2yh + h^2 + 4xy + 4xh - y^2 - 4xy}{h}$$

$$= \lim_{h\to0} \frac{2yh + h^2 + 4xh}{h}$$

$$= \lim_{h\to0} (2y + h + 4x)$$

$$= 4x + 2y$$

17. y could represent the distance from the manufacturing facility, or it could represent the number of components being shipped.

19. y could represent the height (or length) of the mammal.

21. Sketching the graph: 23. Sketching the graph:

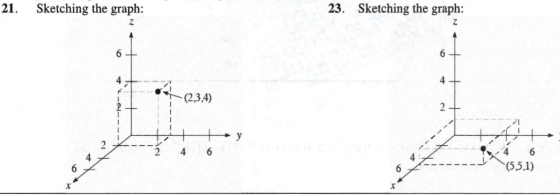

25.   Sketching the graph:

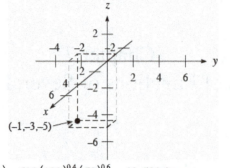

27.   **a.**   Finding the value: $f(120,50) = 600(120)^{0.4}(50)^{0.6} \approx 42,580$ items

b.   Finding the change in value:
$$f(121,50) - f(120,50) = 600(121)^{0.4}(50)^{0.6} - 600(120)^{0.4}(50)^{0.6} \approx 42,722 - 42,580 \approx 142 \text{ items}$$
The change will result in an increase of 142 items produced.

c.   Finding the change in value:
$$f(120,45) - f(120,50) = 600(120)^{0.4}(45)^{0.6} - 600(120)^{0.4}(50)^{0.6} \approx 39,972 - 42,580 \approx -2,608 \text{ items}$$
The change will result in a decrease of 2,608 items produced.

d.   Substituting triple values: $f(360,150) = 600(360)^{0.4}(150)^{0.6} \approx 127,740 = 3 \cdot 42,580$ items
If both labor and capital are tripled, the production will also be tripled.

29.   **a.**   Finding the value: $D(7,6,0) = 10(7)^{2/3} + 8(6)^{4/3} - 7(0)^{3/2} - \ln(7^2 + 6^2 - 0^2) \approx 119$ units

b.   Finding the change in value:
$$D(7,6,5) - D(7,6,0) = \left[10(7)^{2/3} + 8(6)^{4/3} - 7(5)^{3/2} - \ln(7^2 + 6^2 - 5^2) \right]$$
$$- \left[10(7)^{2/3} + 8(6)^{4/3} - 7(0)^{3/2} - \ln(7^2 + 6^2 - 0^2) \right]$$
$$\approx 41 - 119$$
$$\approx -78 \text{ units}$$
The change will result in a decrease of 78 units.

31.   **a.**   Finding the commission: $C(1500,48.50) = 30 + 0.02(1500) + 0.001(1500)(48.50) = \$132.75$

b.   Sketching the graph:

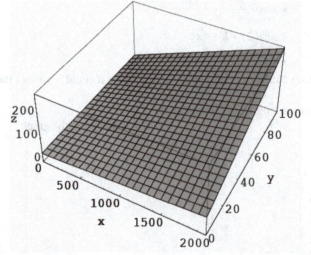

33.   Finding the index: $M(70,45) = -2.5(70)^2 - 3.0(45)^2 + 205(70) + 217(45) + 1.6(70)(45) - 10,000 = 830$
The motivation index level is 830 units.

35. $f(x) = \ln x$ increases at a decreasing rate. $f(x,y) = \ln(xy)$ increases at a decreasing rate in both the x and y directions.

37. Finding the derivative: $f'(y) = 10y - 2$ **39.** Finding the derivative: $\dfrac{df}{dy} = 24k^2y^2 + 20y^3$

41. Finding the derivative: $f'(y) = 4\left(5k^3 - 8y^2\right)^3(-16y) = -64y\left(5k^3 - 8y^2\right)^3$

43. Finding the derivative: $\dfrac{df}{dy} = 0 + \dfrac{5}{jky} \cdot jk = \dfrac{5}{y}$ **45.** $f(x)$ is increasing at a decreasing rate.

7.2 Partial Derivatives

1. **a.** Finding the derivative: $\dfrac{\partial f}{\partial x} = 10x + 6y$

    b. Finding the derivative: $\dfrac{\partial f}{\partial y} = 6x + 24y^2$

    c. Evaluating the derivative: $\dfrac{\partial f}{\partial x}(2,1) = 10(2) + 6(1) = 26$

    d. Evaluating the derivative: $\dfrac{\partial f}{\partial y}(4,2) = 6(4) + 24(2)^2 = 120$

3. **a.** Finding the derivative: $f_x = 3x^2 + 9x^2y^2$

    b. Finding the derivative: $f_y = -8y + 6x^3y$

    c. Evaluating the derivative: $f_x(0,-1) = 3(0)^2 + 9(0)^2(-1)^2 = 0$

    d. Evaluating the derivative: $f_y(-2,1) = -8(1) + 6(-2)^3(1) = -56$

5. **a.** Finding the derivative: $f_x = e^{x+y}$

    b. Finding the derivative: $f_y = e^{x+y}$

    c. Evaluating the derivative: $f_x(1,1) = e^{1+1} = e^2 \approx 7.389$

    d. Evaluating the derivative: $f_y(2,1) = e^{2+1} = e^3 \approx 20.086$

7. **a.** Finding the derivative: $f_x = \dfrac{1}{2 + 4x^2y^2} \cdot 8xy^2 = \dfrac{4xy^2}{1 + 2x^2y^2}$

    b. Finding the derivative: $f_y = \dfrac{1}{2 + 4x^2y^2} \cdot 8x^2y = \dfrac{4x^2y}{1 + 2x^2y^2}$

    c. Evaluating the derivative: $f_x(0,0) = \dfrac{4(0)(0)^2}{1 + 2(0)^2(0)^2} = 0$

    d. Evaluating the derivative: $f_y(0,0) = \dfrac{4(0)^2(0)}{1 + 2(0)^2(0)^2} = 0$

9.

a. Finding the derivative: $f_x = \dfrac{\left(5x - 3y^2\right)(2x) - \left(x^2 + 3y\right)(5)}{\left(5x - 3y^2\right)^2} = \dfrac{10x^2 - 6xy^2 - 5x^2 - 15y}{\left(5x - 3y^2\right)^2} = \dfrac{5x^2 - 6xy^2 - 15y}{\left(5x - 3y^2\right)^2}$

b. Finding the derivative: $f_y = \dfrac{\left(5x - 3y^2\right)(3) - \left(x^2 + 3y\right)(-6y)}{\left(5x - 3y^2\right)^2} = \dfrac{15x - 9y^2 + 6x^2y + 18y^2}{\left(5x - 3y^2\right)^2} = \dfrac{15x + 9y^2 + 6x^2y}{\left(5x - 3y^2\right)^2}$

c. Evaluating the derivative: $f_x(0,2) = \dfrac{5(0)^2 - 6(0)(2)^2 - 15(2)}{\left(5(0) - 3(2)^2\right)^2} = -\dfrac{30}{144} = -\dfrac{5}{24}$

d. Evaluating the derivative: $f_y(0,1) = \dfrac{15(0) + 9(1)^2 + 6(0)^2(1)}{\left(5(0) - 3(1)^2\right)^2} = \dfrac{9}{9} = 1$

11.

a. Finding the derivative: $f_x = 2xe^{3y}$

b. Finding the derivative: $f_y = 3x^2e^{3y}$

c. Evaluating the derivative: $f_x(0,0) = 2(0)e^{3(0)} = 0$

d. Evaluating the derivative: $f_y(0,0) = 3(0)^2 e^{3(0)} = 0$

13. We need to find the first partial derivatives:
$$f_x = 4x - 3 \qquad\qquad f_y = 10y - 4$$

a. Finding the derivative: $f_{xx} = 4$ **b.** Finding the derivative: $f_{yy} = 10$

c. Finding the derivative: $f_{xy} = 0$ **d.** Finding the derivative: $f_{yx} = 0$

15. We need to find the first partial derivatives:
$$f_x = 20x + 4y \qquad\qquad f_y = 10y + 4x$$

a. Finding the derivative: $f_{xx} = 20$ **b.** Finding the derivative: $f_{yy} = 10$

c. Finding the derivative: $f_{xy} = 4$ **d.** Finding the derivative: $f_{yx} = 4$

17. We need to find the first partial derivatives:
$$f_x = 18x - 6y \qquad\qquad f_y = -6x + 2y$$

a. Finding the derivative: $f_{xx} = 18$ **b.** Finding the derivative: $f_{yy} = 2$

c. Finding the derivative: $f_{xy} = -6$ **d.** Finding the derivative: $f_{yx} = -6$

19. We need to find the first partial derivatives:
$$f_x = 16ye^{2x} \qquad\qquad f_y = 8e^{2x}$$

a. Finding the derivative: $f_{xx} = 32ye^{2x}$ **b.** Finding the derivative: $f_{yy} = 0$

c. Finding the derivative: $f_{xy} = 16e^{2x}$ **d.** Finding the derivative: $f_{yx} = 16e^{2x}$

21. We need to find the first partial derivatives:
$$f_x = -6y \qquad\qquad f_y = 18y^2 - 6x \qquad\qquad f_z = -2z$$

a. Finding the derivative: $f_{xx} = 0$ **b.** Finding the derivative: $f_{yy} = 36y$

c. Finding the derivative: $f_z = -2z$ **d.** Finding the derivative: $f_{zy} = 0$

23. Finding the first partial derivatives:
$$f_x = 2x + 3 \qquad\qquad f_y = 14y + 4$$

For both of these to equal 0, we solve the system of equations:
$$2x + 3 = 0 \qquad\qquad\qquad 14y + 4 = 0$$
$$x = -\dfrac{3}{2} \qquad\qquad\qquad\qquad y = -\dfrac{2}{7}$$

The solution is $\left(-\dfrac{3}{2}, -\dfrac{2}{7}\right)$.

25. Finding the first partial derivatives:

$$f_x = 3 - 3x^2 \qquad\qquad f_y = 2 - 3y^2$$

For both of these to equal 0, we solve the system of equations:

$$2 - 3y^2 = 0$$

$$3 - 3x^2 = 0 \qquad\qquad 3y^2 = 2$$
$$3x^2 = 3 \qquad\qquad$$
$$x^2 = 1 \qquad\qquad y^2 = \frac{2}{3}$$
$$x = -1, 1 \qquad\qquad$$
$$y = -\sqrt{\frac{2}{3}}, \sqrt{\frac{2}{3}}$$

The solutions are $\left(-1, -\sqrt{\frac{2}{3}}\right), \left(-1, \sqrt{\frac{2}{3}}\right), \left(1, -\sqrt{\frac{2}{3}}\right), \left(1, \sqrt{\frac{2}{3}}\right)$.

27. At the point $(100,600)$, if x increases from 100 to 101, $N(x,y)$ will increase by approximately 30. At the point $(80,420)$, if y increases from 420 to 421, $N(x,y)$ will increase by approximately 16.

29. At the point $(2,1,3,3,4)$, if y increases from 1 to 2, $S(x,y,z,w,m)$ will decrease by approximately 65.

31. If the country has \$40 million in World Bank loans and a population of 70 million people, if the population increases the acres of rainforest destroyed increases at an increasing rate.

33. If a persons chronological age increases, their IQ decreases at a decreasing rate.

35. As the age of the person increases, the time to recover from pneumonia increases at a decreasing rate.

37. As the concentration of residents in public housing increases, the number of arsons in the city increases at an increasing rate.

39. Evaluating the derivatives:

$$\frac{\partial f}{\partial x}(20,40) = 0.3(20)^{-0.7}(40)^{0.7} \approx 0.49$$

$$\frac{\partial^2 f}{\partial x^2}(20,40) = -0.21(20)^{-1.7}(40)^{0.7} \approx -0.02$$

As labor is increased from 20 to 21 units, the production will increase by approximately 0.49 units but this rate will decrease by approximately 0.02 units.

41. The demand for printers decreases as the price of the printer increases, and the demand for printers decreases as the price of the ink cartridges increases.

43. a. Evaluating the function: $R(20,15) = 350(20) + 600(15) - 4(20)^2 - 3(15)^2 = 13,725$

   The revenue is \$13,725 if 20 units of product A and 15 units of product B are sold.

   b. Finding the derivative: $\dfrac{\partial R}{\partial x} = 350 - 8x$

   Evaluating the derivative: $\dfrac{\partial R}{\partial x}(20,15) = 350 - 8(20) = 190$

   The revenue will increase by \$190 if x increases from 20 to 21 units.

   c. Finding the derivative: $\dfrac{\partial R}{\partial y} = 600 - 6y$

   Evaluating the derivative: $\dfrac{\partial R}{\partial y}(20,15) = 600 - 6(15) = 510$

   The revenue will decrease by \$510 if y decreases from 15 to 14 units.

45. **a.** Evaluating the function: $C(40,22,36) = 2.5(40-22)(36)^{-0.67} \approx 4.08$

The oxygen consumption is 4.08 of an animal weighing 36 kg with an internal temperature of 40°C and a fur temperature of 22°C.

b. Finding the derivative: $\dfrac{\partial C}{\partial F} = -2.5W^{-0.67}$

Evaluating the derivative: $\dfrac{\partial C}{\partial F}(40,22,36) = -2.5(36)^{-0.67} \approx -0.23$

The oxygen consumption will decrease by 0.23 if the fur temperature increases from 22°C to 23°C.

c. Finding the derivative: $\dfrac{\partial C}{\partial W} = -1.675(T-F)W^{-1.67}$

Evaluating the derivative: $\dfrac{\partial C}{\partial W}(40,22,36) = -1.675(40-22)(36)^{-1.67} \approx -0.08$

The oxygen consumption will decrease by 0.08 if the animal's weight decreases from 36 kg to 35 kg.

47. **a.** Evaluating the function: $f(27,64) = 75(27)^{2/3}(64)^{1/3} = 2{,}700$

Thus 2,700 units are produced if 27 units of labor and 64 units of capital are used.

b. Finding the derivative: $\dfrac{\partial f}{\partial x} = 50x^{-1/3}y^{1/3}$

Evaluating the derivative: $\dfrac{\partial f}{\partial x}(27,64) = 50(27)^{-1/3}(64)^{1/3} \approx 66.67$

The production will increase by 66.67 units if the units of labor increase from 27 to 28 units.

c. Finding the derivative: $\dfrac{\partial f}{\partial y} = 25x^{2/3}y^{-2/3}$

Evaluating the derivative: $\dfrac{\partial f}{\partial y}(27,64) = 25(27)^{2/3}(64)^{-2/3} \approx 14.06$

The production will decrease by 14.06 units if the units of capital decrease from 64 to 63 units.

49. Finding the required values:

$$f(1,0) = e^{-(1)(0)} + e^{0/1} = 2$$

$$f_x(x,y) = -ye^{-xy} - \frac{y}{x^2}e^{y/x}$$

$$f_y(x,y) = -xe^{-xy} + \frac{1}{x}e^{y/x}$$

$$f_x(1,0) = -(0)e^{-(1)(0)} - \frac{0}{(1)^2}e^{0/1} = 0$$

$$f_y(1,0) = -(1)e^{-(1)(0)} + \frac{1}{1}e^{0/1} = 0$$

51. Finding and evaluating the derivatives:

$$\frac{\partial f}{\partial x} = 15.05x^{-0.57}y^{0.57}$$

$$\frac{\partial f}{\partial x}(15,21) = 15.05(15)^{-0.57}(21)^{0.57} \approx 18.23$$

$$\frac{\partial f}{\partial y} = 19.95x^{0.43}y^{-0.43}$$

$$\frac{\partial f}{\partial y}(15,21) = 19.95(15)^{0.43}(21)^{-0.43} \approx 17.26$$

When 15 units of labor and 21 units of capital are used, an increase of 1 unit in labor will result in an increase of 18.23 units of production, and an increase of 1 unit in capital will result in an increase of 17.26 units of production.

53. Finding the derivatives:
$$f_x(x,y) = 3x + 6 \qquad\qquad f_y(x,y) = 2y - 8$$

55. Finding the derivatives:
$$f_x(x,y) = 3x \qquad\qquad f_y(x,y) = 2y^3 - 2y$$

57. Finding the derivatives:
$$f_x(x,y) = 4x + 8 \qquad f_y(x,y) = 6y - 12$$
$$f_{xx}(x,y) = 4 \qquad f_{yy}(x,y) = 6 \qquad\qquad f_{xy}(x,y) = 0$$

Now evaluating the expression:
$$f_{xx}(x,y) \cdot f_{yy}(x,y) - f_{xy}^2(x,y) = 4 \cdot 6 - (0)^2 = 24$$

59. Finding the derivatives:
$$f_x(x,y) = 2x^2 - 50 \qquad\qquad f_y(x,y) = 4y^2 - 16y$$

Solving the system of equations:
$$2x^2 - 50 = 0$$
$$2x^2 = 50 \qquad\qquad 4y^2 - 16y = 0$$
$$x^2 = 25 \qquad\qquad 4y(y - 4) = 0$$
$$x = -5, 5 \qquad\qquad y = 0, 4$$

The solutions are $(-5,0)$, $(-5,4)$, $(5,0)$, and $(5,4)$.

7.3 Optimization of Functions of Two Variables

1. Finding the first and second derivatives:
$$f_x(x,y) = 3x + 15 \qquad\qquad f_y(x,y) = 2y - 8$$
$$f_{xx}(x,y) = 3 \qquad\qquad f_{yy}(x,y) = 2 \qquad\qquad f_{xy}(x,y) = 0$$

Finding the critical points by setting the first derivatives equal to 0:
$$3x + 15 = 0 \qquad\qquad 2y - 8 = 0$$
$$3x = -15 \qquad\qquad 2y = 8$$
$$x = -5 \qquad\qquad y = 4$$

Now evaluating the expression: $D(x,y) = f_{xx}(x,y) \cdot f_{yy}(x,y) - f_{xy}^2(x,y) = (3)(2) - (0)^2 = 6$

Since $D(-5,4) = 6 > 0$ and $f_{xx}(-5,4) = 3 > 0$, $(-5,4)$ is a relative minimum.

3. Finding the first and second derivatives:
$$f_x(x,y) = -4x - 12 \qquad\qquad f_y(x,y) = 5y - 20$$
$$f_{xx}(x,y) = -4 \qquad\qquad f_{yy}(x,y) = 5 \qquad\qquad f_{xy}(x,y) = 0$$

Finding the critical points by setting the first derivatives equal to 0:
$$-4x - 12 = 0 \qquad\qquad 5y - 20 = 0$$
$$-4x = 12 \qquad\qquad 5y = 20$$
$$x = -3 \qquad\qquad y = 4$$

Now evaluating the expression: $D(x,y) = f_{xx}(x,y) \cdot f_{yy}(x,y) - f_{xy}^2(x,y) = (-4)(5) - (0)^2 = -20$

Since $D(-3,4) = -20 < 0$, $(-3,4)$ is a saddle point.

5. Finding the first and second derivatives:

$$f_x(x,y) = -2x - y + 1 \qquad\qquad f_y(x,y) = -2y - x + 6$$

$$f_{xx}(x,y) = -2 \qquad\qquad f_{yy}(x,y) = -2 \qquad\qquad f_{xy}(x,y) = -1$$

Finding the critical points by setting the first derivatives equal to 0 results in the system:

$$-2x - y = -1$$
$$-x - 2y = -6$$

Multiplying the first equation by –2:

$$4x + 2y = 2$$
$$-x - 2y = -6$$

Adding yields:

$$3x = -4$$

$$x = -\frac{4}{3}$$

Substituting into the first equation:

$$-2\left(-\frac{4}{3}\right) - y = -1$$

$$\frac{8}{3} - y = -1$$

$$-y = -\frac{11}{3}$$

$$y = \frac{11}{3}$$

Now evaluating the expression: $D(x,y) = f_{xx}(x,y) \cdot f_{yy}(x,y) - f_{xy}^2(x,y) = (-2)(-2) - (-1)^2 = 3$

Since $D\left(-\frac{4}{3}, \frac{11}{3}\right) = 3 > 0$ and $f_{xx}\left(-\frac{4}{3}, \frac{11}{3}\right) = -2 < 0$, $\left(-\frac{4}{3}, \frac{11}{3}\right)$ is a relative maximum.

7. Finding the first and second derivatives:

$$f_x(x,y) = 3x^2 - 3 \qquad\qquad f_y(x,y) = -2y + 4$$

$$f_{xx}(x,y) = 6x \qquad\qquad f_{yy}(x,y) = -2 \qquad\qquad f_{xy}(x,y) = 0$$

Finding the critical points by setting the first derivatives equal to 0 results in the equations:

$$3x^2 - 3 = 0 \qquad\qquad\qquad -2y + 4 = 0$$
$$x^2 = 1 \qquad\qquad\qquad\qquad -2y = -4$$
$$x = -1, 1 \qquad\qquad\qquad\qquad y = 2$$

Now evaluating the expression: $D(x,y) = f_{xx}(x,y) \cdot f_{yy}(x,y) - f_{xy}^2(x,y) = (6x)(-2) - (0)^2 = -12x$

Since $D(-1,2) = -12(-1) = 12 > 0$ and $f_{xx}(-1,2) = 6(-1) = -6 < 0$, $(-1,2)$ is a relative maximum.

Since $D(1,2) = -12(1) = -12 < 0$, $(1,2)$ is a saddle point.

9. Finding the first and second derivatives:

$$f_x(x,y) = 4x \qquad\qquad f_y(x,y) = 8y^3 + 1$$

$$f_{xx}(x,y) = 4 \qquad\qquad f_{yy}(x,y) = 24y^2 \qquad\qquad f_{xy}(x,y) = 0$$

Finding the critical points by setting the first derivatives equal to 0 results in the equations:

$$8y^3 + 1 = 0$$

$$4x = 0$$

$$x = 0$$

$$y^3 = -\frac{1}{8}$$

$$y = -\frac{1}{2}$$

Now evaluating the expression: $D(x,y) = f_{xx}(x,y) \cdot f_{yy}(x,y) - f_{xy}^2(x,y) = (4)(24y^2) - (0)^2 = 96y^2$

Since $D\left(0,-\frac{1}{2}\right) = 96\left(-\frac{1}{2}\right)^2 = 24 > 0$ and $f_{xx}\left(0,-\frac{1}{2}\right) = 4 > 0$, $\left(0,-\frac{1}{2}\right)$ is a relative minimum.

11. Finding the first and second derivatives:

$$f_x(x,y) = 3x^2 - 3y \qquad\qquad f_y(x,y) = -3x + 3y^2$$

$$f_{xx}(x,y) = 6x \qquad\qquad f_{yy}(x,y) = 6y \qquad\qquad f_{xy}(x,y) = -3$$

Finding the critical points by setting the first derivatives equal to 0 results in the equations:

$$3x^2 - 3y = 0 \qquad\qquad\qquad -3x + 3y^2 = 0$$

$$3x^2 = 3y \qquad\qquad\qquad 3x = 3y^2$$

$$y = x^2 \qquad\qquad\qquad x = y^2$$

Substituting:

$$x = \left(x^2\right)^2$$

$$x = x^4$$

$$x^4 - x = 0$$

$$x\left(x^3 - 1\right) = 0$$

$$x = 0, 1$$

$$y = 0, 1$$

Now evaluating the expression:

$$D(x,y) = f_{xx}(x,y) \cdot f_{yy}(x,y) - f_{xy}^2(x,y) = (6x)(6y) - (-3)^2 = 36xy - 9$$

Since $D(0,0) = 36(0)(0) - 9 = -9 < 0$, $(0,0)$ is a saddle point.

Since $D(1,1) = 36(1)(1) - 9 = 27 > 0$ and $f_{xx}(1,1) = 6(1) = 6 > 0$, $(1,1)$ is a relative minimum.

13. Finding the first and second derivatives:

$$f_x(x,y) = \frac{2}{3}x^{-1/3} = \frac{2}{3x^{1/3}} \qquad\qquad f_y(x,y) = \frac{2}{3}y^{-1/3} = \frac{2}{3y^{1/3}}$$

$$f_{xx}(x,y) = -\frac{2}{9}x^{-4/3} = -\frac{2}{9x^{4/3}} \qquad f_{yy}(x,y) = -\frac{2}{9}y^{-4/3} = -\frac{2}{9y^{4/3}} \qquad f_{xy}(x,y) = 0$$

Finding the critical points by setting the first derivatives equal to 0 results in no solution. There are no critical points.

15. Evaluating the expression: $D(x,y) = f_{xx}(x,y) \cdot f_{yy}(x,y) - f_{xy}^2(x,y) = (5)(3) - (0)^2 = 15$

Since $D(3,-2) = 15 > 0$ and $f_{xx}(3,-2) = 5 > 0$, $(3,-2)$ is a relative minimum.

17. Evaluating the expression: $D(x,y) = f_{xx}(x,y) \cdot f_{yy}(x,y) - f_{xy}^2(x,y) = (-8)(1) - (0)^2 = -8$

Since $D(2,-6) = -8 < 0$, $(2,-6)$ is a saddle point.

19. Evaluating the expression: $D(x,y) = f_{xx}(x,y) \cdot f_{yy}(x,y) - f_{xy}^2(x,y) = (-2)(2) - (8)^2 = -68$

Since $D(-1,0) = -68 < 0$, $(3,5)$ is a saddle point.

21. Finding the first and second derivatives:

$R_x(x,y) = -0.14x + 4 + 2y$ $R_y(x,y) = -200y + 5 + 2x$

$R_{xx}(x,y) = -0.14$ $R_{yy}(x,y) = -200$ $R_{xy}(x,y) = 2$

Finding the critical points by setting the first derivatives equal to 0 results in the system:

$$-0.14x + 2y = -4$$
$$2x - 200y = -5$$

Multiplying the first equation by 100:

$$-14x + 200y = -400$$
$$2x - 200y = -5$$

Adding yields:

$$-12x = -405$$
$$x = 33.75$$

Substituting into the first equation:

$$-0.14(33.75) + 2y = -4$$
$$-4.725 + 2y = -4$$
$$2y = 0.725$$
$$y = 0.3625$$

Now evaluating the expression: $D(x,y) = R_{xx}(x,y) \cdot R_{yy}(x,y) - R_{xy}^2(x,y) = (-0.14)(-200) - (2)^2 = 24$

Since $D(33.75, 0.3625) = 24 > 0$ and $R_{xx}(33.75, 0.3625) = -0.14 < 0$, $(33.75, 0.3625)$ is a relative maximum.

The company will maximize its revenue by spending \$33,750 on radio advertisements and \$362.50 on newspaper advertisements.

23. Using x, y, and z as the dimensions, then:

$$xyz = 64$$
$$z = \frac{64}{xy}$$

The surface area is then given by the function: $A(x,y) = 2xy + 2xz + 2yz = 2xy + 2x\left(\frac{64}{xy}\right) + 2y\left(\frac{64}{xy}\right) = 2xy + \frac{128}{y} + \frac{128}{x}$

Finding the first and second derivatives:

$A_x(x,y) = 2y - \dfrac{128}{x^2}$ $A_y(x,y) = 2x - \dfrac{128}{y^2}$

$A_{xx}(x,y) = \dfrac{256}{x^3}$ $A_{yy}(x,y) = \dfrac{256}{y^3}$ $A_{xy}(x,y) = 2$

Finding the critical points by setting the first derivatives equal to 0 results in the equations:

$2y - \dfrac{128}{x^2} = 0$ $2x - \dfrac{128}{y^2} = 0$

$2y = \dfrac{128}{x^2}$ $2x = \dfrac{128}{y^2}$

$y = \dfrac{64}{x^2}$ $x = \dfrac{64}{y^2}$

Substituting:

$$x = \frac{64}{\left(\dfrac{64}{x^2}\right)^2}$$

$$x = \frac{x^4}{64}$$

$$x^4 - 64x = 0$$

$$x\left(x^3 - 64\right) = 0$$

$$x = 4 \qquad (x \neq 0)$$

$$y = 4$$

Now evaluating the expression: $D(x,y) = A_{xx}(x,y) \cdot A_{yy}(x,y) - A_{xy}^2(x,y) = \left(\dfrac{256}{x^3}\right)\left(\dfrac{256}{y^3}\right) - (2)^2 = \dfrac{65{,}536}{x^3 y^3} - 4$

Since $D(4,4) = \dfrac{65{,}536}{(4)^3 (4)^3} - 4 = 12 > 0$ and $A_{xx}(4,4) = \dfrac{256}{4^3} = 4 > 0$, $(4,4)$ is a relative minimum.

The box dimensions should be 4 in. by 4 in. by 4 in.

25. Let x and y represent the width and height, and let l represent the length, then:

$$2x + 2y + l = 84$$

$$l = 84 - 2x - 2y$$

The volume is then given by the function: $V(x,y) = xyl = xy(84 - 2x - 2y) = 84xy - 2x^2 y - 2xy^2$

Finding the first and second derivatives:

$$V_x(x,y) = 84y - 4xy - 2y^2 \qquad V_y(x,y) = 84x - 2x^2 - 4xy$$

$$V_{xx}(x,y) = -4y \qquad\qquad V_{yy}(x,y) = -4x \qquad\qquad V_{xy}(x,y) = 84 - 4x - 4y$$

Finding the critical points by setting the first derivatives equal to 0 results in the equations:

$$84y - 4xy - 2y^2 = 0 \qquad\qquad\qquad 84x - 2x^2 - 4xy = 0$$

$$2y(42 - 2x - y) = 0 \qquad\qquad\qquad 2x(42 - x - 2y) = 0$$

$$42 - 2x - y = 0 \qquad\qquad\qquad\qquad 42 - x - 2y = 0$$

This yields the system of equations:

$$2x + y = 42$$

$$x + 2y = 42$$

Multiply the second equation by –2:

$$2x + y = 42$$

$$-2x - 4y = -84$$

Adding:

$$-3y = -42$$

$$y = 14$$

Substituting:

$$2x + 14 = 42$$

$$2x = 28$$

$$x = 14$$

Now evaluating the expression: $D(x,y) = V_{xx}(x,y) \cdot V_{yy}(x,y) - V_{xy}^2(x,y) = (-4y)(-4x) - (84 - 4x - 4y)^2$

Since $D(14,14) = (-56)(-56) - (84 - 56 - 56)^2 = 2{,}352 > 0$ and $V_{xx}(14,14) = -4(14) = -56 < 0$, $(14,14)$ is a relative maximum. The package dimensions should be 14 in. by 14 in. by 28 in.

27.   **a.**   Using Wolfram|Alpha, the relative maximum is at $(0,0,300)$.
       b.   Using Wolfram|Alpha, the relative maximum is at $(0,0,3)$.
       c.   Using Wolfram|Alpha, the relative maximum is at $(0,0,1)$.
       d.   Using Wolfram|Alpha, the relative maximum is at $(0,0,300/n)$.

29.   Finding the first and second derivatives:

$$P_x(x,y) = 22 - 2.4x \qquad\qquad P_y(x,y) = 73 - 3y$$

$$P_{xx}(x,y) = 22 \qquad\qquad P_{yy}(x,y) = 73 \qquad\qquad P_{xy}(x,y) = 0$$

Finding the critical points by setting the first derivatives equal to 0 results in the equations:

$$22 - 2.4x = 0 \qquad\qquad\qquad\qquad 73 - 3y = 0$$

$$2.4x = 22 \qquad\qquad\qquad\qquad 3y = 73$$

$$x = \frac{55}{6} \approx 9.167 \qquad\qquad\qquad\qquad y = \frac{73}{3} \approx 24.333$$

Now evaluating the expression: $D(x,y) = P_{xx}(x,y) \cdot P_{yy}(x,y) - P_{xy}^2(x,y) = (22)(73) - (0)^2 = 1{,}606$

Since $D\left(\dfrac{55}{6}, \dfrac{73}{3}\right) = 1{,}606 > 0$ and $P_{xx}\left(\dfrac{55}{6}, \dfrac{73}{3}\right) = 22 > 0$, $(9.167, 24.333, 989)$ is a relative minimum.

Approximately 9 units of Product A and 24 units of Product B should be produced to maximize the profit. The maximum profit will be \$989.

31.   Solving for x yields $x = 2y + 4$, now substitute:

$$f(y) = (2y+4)^2 - 3y^2 + 2(2y+4) + 4y = 4y^2 + 16y + 16 - 3y^2 + 4y + 8 + 4y = y^2 + 24y + 24$$

33.   Finding the partial derivatives:

$$F_x(x,y,\lambda) = 2x + \lambda$$

$$F_y(x,y,\lambda) = 20y - \lambda$$

$$F_\lambda(x,y,\lambda) = x - y - 18$$

35.   Adding the first two equations results in the system:

$$2x + 20y = 0$$

$$x - y = 18$$

Multiplying the second equation by –2:

$$2x + 20y = 0$$

$$-2x + 2y = -36$$

Adding yields:

$$22y = -36$$

$$y = -\frac{18}{11}$$

Substituting into the third equation:

$$x - \left(-\frac{18}{11}\right) = 18$$

$$x + \frac{18}{11} = 18$$

$$x = \frac{180}{11}$$

Substituting into the first equation:

$$2\left(\frac{180}{11}\right) + \lambda = 0$$

$$\lambda = -\frac{360}{11}$$

The solution is $x = \frac{180}{11}$, $y = -\frac{18}{11}$, $\lambda = -\frac{360}{11}$.

7.4 Constrained Maxima and Minima

1. Solving for x yields $x = y + 7$, now substitute:

$$f(y) = (y+7)^2 - 8y^2 + 5(y+7) + 9y + 8 = y^2 + 14y + 49 - 8y^2 + 5y + 35 + 9y + 8 = -7y^2 + 28y + 43$$

Differentiating:

$$f'(y) = 0$$
$$-14y + 28 = 0$$
$$-14y = -28$$
$$y = 2$$
$$x = 2 + 7 = 9$$

Since $f''(y) = -14 < 0$, the curve is concave down resulting in a maximum point. Evaluating the function:

$$f(9,2) = (9)^2 - 8(2)^2 + 5(9) + 9(2) + 8 = 120$$

The relative maximum is (9,2,120).

3. Solving for x yields $x = 5 - 4y$, now substitute:

$$f(y) = (5-4y)^2 + y^2 + 3(5-4y)y + 7(5-4y) + 63y - 60$$
$$= 25 - 40y + 16y^2 + y^2 + 15y - 12y^2 + 35 - 28y + 63y - 60$$
$$= 5y^2 + 10y$$

Differentiating:

$$f'(y) = 0$$
$$10y + 10 = 0$$
$$10y = -10$$
$$y = -1$$
$$x = 5 - 4(-1) = 9$$

Since $f''(y) = 10 > 0$, the curve is concave up resulting in a minimum point. Evaluating the function:

$$f(9,-1) = (9)^2 + (-1)^2 + 3(9)(-1) + 7(9) + 63(-1) - 60 = -5$$

The relative minimum is (9,–1,–5).

5. Construct the Lagrangian function:
$$F(x,y,\lambda) = f(x,y) + \lambda g(x,y) = 25 - x^2 - y^2 + \lambda(x + y + 1)$$
Finding the first-order partial derivatives and setting them equal to 0:
$$F_x = -2x + \lambda = 0$$
$$F_y = -2y + \lambda = 0$$
$$F_\lambda = x + y + 1 = 0$$
Solving for λ in the first two equations yields:
$$\lambda = 2x$$
$$\lambda = 2y$$
Therefore:
$$2x = 2y$$
$$x = y$$
Substituting into $x + y + 1 = 0$:
$$y + y + 1 = 0$$
$$2y + 1 = 0$$
$$2y = -1$$
$$y = -\frac{1}{2}$$
$$x = -\frac{1}{2}$$

Evaluating the function: $f\left(-\dfrac{1}{2}, -\dfrac{1}{2}\right) = 25 - \left(-\dfrac{1}{2}\right)^2 - \left(-\dfrac{1}{2}\right)^2 = 25 - \dfrac{1}{4} - \dfrac{1}{4} = 25 - \dfrac{1}{2} = \dfrac{49}{2}$

7. Construct the Lagrangian function:
$$F(x,y,\lambda) = f(x,y) + \lambda g(x,y) = y^2 + 6x + \lambda(y - 2x)$$
Finding the first-order partial derivatives and setting them equal to 0:
$$F_x = 6 - 2\lambda = 0$$
$$F_y = 2y + \lambda = 0$$
$$F_\lambda = y - 2x = 0$$
Solving for λ in the first two equations yields:
$$\lambda = 3$$
$$2y + 3 = 0$$
Therefore:
$$2y + 3 = 0$$
$$2y = -3$$
$$y = -\frac{3}{2}$$
Substituting into $y - 2x = 0$:
$$-\frac{3}{2} - 2x = 0$$
$$-2x = \frac{3}{2}$$
$$x = -\frac{3}{4}$$

Evaluating the function: $f\left(-\dfrac{3}{4}, -\dfrac{3}{2}\right) = \left(-\dfrac{3}{2}\right)^2 + 6\left(-\dfrac{3}{4}\right) = \dfrac{9}{4} - \dfrac{9}{2} = -\dfrac{9}{4}$

9. Construct the Lagrangian function:
$$F(x,y,\lambda) = f(x,y) + \lambda g(x,y) = 6x^{1/3}y^{2/3} + \lambda(4x + 3y - 36)$$
Finding the first-order partial derivatives and setting them equal to 0:
$$F_x = 2x^{-2/3}y^{2/3} + 4\lambda = 0$$
$$F_y = 4x^{1/3}y^{-1/3} + 3\lambda = 0$$
$$F_\lambda = 4x + 3y - 36 = 0$$
Solving for λ in the first two equations yields:
$$\lambda = -\frac{x^{-2/3}y^{2/3}}{2}$$
$$\lambda = -\frac{4x^{1/3}y^{-1/3}}{3}$$
Therefore:
$$-\frac{x^{-2/3}y^{2/3}}{2} = -\frac{4x^{1/3}y^{-1/3}}{3}$$
$$\frac{y^{2/3}}{2x^{2/3}} = \frac{4x^{1/3}}{3y^{1/3}}$$
$$3y = 8x$$
$$y = \frac{8}{3}x$$
Substituting into $4x + 3y - 36 = 0$:
$$4x + 3\left(\frac{8}{3}x\right) - 36 = 0$$
$$12x - 36 = 0$$
$$12x = 36$$
$$x = 3$$
$$y = \frac{8}{3}(3) = 8$$

Evaluating the function: $f(3,8) = 6(3)^{1/3}(8)^{2/3} = 24\sqrt[3]{3} \approx 34.6$

11. If the course requires 1 more hour of instruction, the achievement level will increase by about 1.3.
13. If the volume of the can is increased by 1 cubic inch, the surface area will increase by about 1.32 square inches.
15. If the number of items tested is increased by 1, the cost will increase by about $4.09.
17. The cost of labor is: ($8 per hour)(8 hours per day)(260 days)$x = 16,640x$
Therefore the constraint equation is: $16,640x + y = 1,622,400$

Write the constraint function as: $g(x,y) = 16,640x + y - 1,622,400 = 0$

Now form the Lagrangian function: $F(x,y,\lambda) = f(x,y) + \lambda g(x,y) = 12x^{2/3}y^{1/3} + \lambda(16,640x + y - 1,622,400)$
Finding the first-order partial derivatives and setting them equal to 0:
$$F_x = 8x^{-1/3}y^{1/3} + 16,640\lambda = 0$$
$$F_y = 4x^{2/3}y^{-2/3} + \lambda = 0$$
$$F_\lambda = 16,640x + y - 1,622,400 = 0$$
Solving for λ in the first two equations yields:
$$\lambda = -\frac{8x^{-1/3}y^{1/3}}{16,640}$$
$$\lambda = -4x^{2/3}y^{-2/3}$$

Therefore:

$$-\frac{8x^{-1/3}y^{1/3}}{16,640} = -4x^{2/3}y^{-2/3}$$

$$\frac{y^{1/3}}{x^{1/3}} = \frac{8,320x^{2/3}}{y^{2/3}}$$

$$y = 8,320x$$

Substituting into $16,640x + y - 1,622,400 = 0$:

$$16,640x + 8,320x - 1,622,400 = 0$$

$$24,960x - 1,622,400 = 0$$

$$24,960x = 1,622,400$$

$$x = 65$$

$$y = 8,320(65) = 540,800$$

$$\lambda = -4(65)^{2/3}(540800)^{-2/3} \approx -0.0097$$

Evaluating the function: $f(65,540800) = 12(65)^{2/3}(540800)^{1/3} \approx 15,805.29$

To maximize production at 15,805.29 units, 65 units (or \$1,081,600) should be allocated to labor and \$540,800 should be allocated to capital. Further, if the number of dollars allocated is increased by \$1, the number of units decreases by 0.0097 units.

19. The cost of labor is: $(\$12 \text{ per hour})(8 \text{ hours per day})(22 \text{ days})x = 2,112x$

Therefore the constraint equation is: $2,112x + y = 48,032$

Write the constraint function as: $g(x,y) = 2,112x + y - 48,032 = 0$

Now form the Lagrangian function: $F(x,y,\lambda) = f(x,y) + \lambda g(x,y) = x^{3/5}y^{2/5} + \lambda(2,112x + y - 48,032)$

Finding the first-order partial derivatives and setting them equal to 0:

$$F_x = \frac{3}{5}x^{-2/5}y^{2/5} + 2,112\lambda = 0$$

$$F_y = \frac{2}{5}x^{3/5}y^{-3/5} + \lambda = 0$$

$$F_\lambda = 2,112x + y - 48,032 = 0$$

Solving for λ in the first two equations yields:

$$\lambda = -\frac{3x^{-2/5}y^{2/5}}{10,560}$$

$$\lambda = -\frac{2x^{3/5}y^{-3/5}}{5}$$

Therefore:

$$-\frac{3x^{-2/5}y^{2/5}}{10,560} = -\frac{2x^{3/5}y^{-3/5}}{5}$$

$$\frac{y^{2/5}}{x^{2/5}} = \frac{1,408x^{3/5}}{y^{3/5}}$$

$$y = 1,408x$$

Substituting into $2,112x + y - 48,032 = 0$:

$$2,112x + 1,408x - 48,032 = 0$$
$$3,520x - 48,032 = 0$$
$$3,520x = 48,032$$
$$x = 13.65$$
$$y = 1,408(13.65) = 19,212.80$$
$$\lambda = \frac{-2(13.65)^{3/5}(19,212.80)^{-3/5}}{5} \approx -0.0052$$

Evaluating the function: $f(13.65, 19212.80) = (13.65)^{3/5}(19,212.80)^{2/5} \approx 248.037$

To maximize production at 248.037 units, 13.65 units (or $28,819.20) should be allocated to labor and $19,212.80 should be allocated to capital. Further, if the number of dollars allocated is increased by $1, the number of units decreases by 0.0052 units.

21. The constraint equation is: $x = 5y$

Write the constraint function as: $g(x,y) = x - 5y = 0$

Now form the Lagrangian function: $F(x,y,\lambda) = C(x,y) + \lambda g(x,y) = 6x^2 + 3y^2 - 612x + 20,000 + \lambda(x - 5y)$

Finding the first-order partial derivatives and setting them equal to 0:

$$F_x = 12x - 612 + \lambda = 0$$
$$F_y = 6y - 5\lambda = 0$$
$$F_\lambda = x - 5y = 0$$

Solving for λ in the first two equations yields:

$$\lambda = 612 - 12x$$
$$\lambda = \frac{6}{5}y$$

Therefore:

$$612 - 12x = \frac{6}{5}y$$
$$y = 510 - 10x$$

Substituting into $x - 5y = 0$:

$$x - 5(510 - 10x) = 0$$
$$x - 2,550 + 50x = 0$$
$$51x = 2,550$$
$$x = 50$$
$$y = 510 - 10(50) = 10$$

Evaluating the function: $C(50,10) = 6(50)^2 + 3(10)^2 - 612(50) + 20,000 = \$4,700$

To minimize cost at $4,700, site A should produce 50 units and site B should produce 10 units.

23. The constraint equation is $100x + 50y = 42{,}000$, or $2x + y = 840$.

Write the constraint function as: $g(x,y) = 2x + y - 840 = 0$

Now form the Lagrangian function: $F(x,y,\lambda) = N(x,y) + \lambda g(x,y) = 30x(4y+40)^{2/3} + \lambda(2x+y-840)$

Finding the first-order partial derivatives and setting them equal to 0:

$$F_x = 30(4y+40)^{2/3} + 2\lambda = 0$$

$$F_y = 20x(4y+40)^{-1/3} \cdot 4 + \lambda = 80x(4y+40)^{-1/3} + \lambda = 0$$

$$F_\lambda = 2x + y - 840 = 0$$

Solving for λ in the first two equations yields:

$$\lambda = -15(4y+40)^{2/3}$$

$$\lambda = -80x(4y+40)^{-1/3}$$

Therefore:

$$-15(4y+40)^{2/3} = -80x(4y+40)^{-1/3}$$

$$4y+40 = \frac{16}{3}x$$

$$4y = \frac{16}{3}x - 40$$

$$y = \frac{4}{3}x - 10$$

Substituting into $2x + y - 840 = 0$:

$$2x + \frac{4}{3}x - 10 - 840 = 0$$

$$\frac{10}{3}x - 850 = 0$$

$$\frac{10}{3}x = 850$$

$$x = 255$$

$$y = \frac{4}{3}(255) - 10 = 330$$

$$\lambda = -15(4 \cdot 330 + 40)^{2/3} \approx -1{,}841$$

Evaluating the function: $N(255,330) = 30(255)(4 \cdot 330 + 40)^{2/3} \approx 939{,}047$

To maximize the number of people reached at 939,047, the franchise should purchase 255 radio and 330 newspaper ads. If the number of ads is increased by 1, the number of people reached will decrease by 1,841 people.

25. The constraint equation is: $x + y = 2,000$

Write the constraint function as: $g(x,y) = x + y - 2,000 = 0$

Now form the Lagrangian function: $F(x,y,\lambda) = f(x,y) + \lambda g(x,y) = 1 - \frac{1}{4}e^{-x/5} - \frac{1}{4}e^{-y/70} + \lambda(x + y - 2,000)$

Finding the first-order partial derivatives and setting them equal to 0:

$$F_x = \frac{1}{20}e^{-x/5} + \lambda = 0$$

$$F_y = \frac{1}{280}e^{-y/70} + \lambda = 0$$

$$F_\lambda = x + y - 2,000 = 0$$

Solving for λ in the first two equations yields:

$$\lambda = -\frac{1}{20}e^{-x/5}$$

$$\lambda = -\frac{1}{280}e^{-y/70}$$

Therefore:

$$-\frac{1}{20}e^{-x/5} = -\frac{1}{280}e^{-y/70}$$

$$e^{-y/70} = 14e^{-x/5}$$

$$-\frac{y}{70} = \ln\left(14e^{-x/5}\right)$$

$$-\frac{y}{70} = \ln 14 - \frac{x}{5}$$

$$y = -70\ln 14 + 14x$$

Substituting into $x + y - 2,000 = 0$:

$$x - 70\ln 14 + 14x - 2,000 = 0$$

$$15x = 2,000 + 70\ln 14$$

$$x \approx 145.65$$

$$y \approx 2,000 - 145.65 = 1,854.35$$

To maximize the percentage of pests killed, the manufacturer should use 145.65 pounds of pesticide A and 1,854.35 pounds of pesticide B.

27. **a.** Sketching the graph: **b.** Sketching the graph:

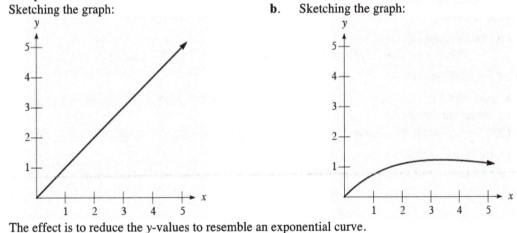

  c. The effect is to reduce the y-values to resemble an exponential curve.

d.   Sketching the graph:

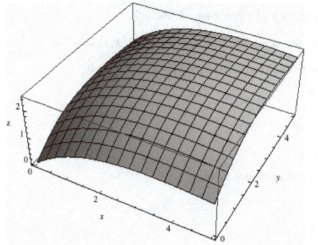

The effect seems to be the same along the y-axis.

e.   Using Wolfram|Alpha, the maximum is approximately 2.45 when $(x,y) = \left(\dfrac{10}{3}, \dfrac{10}{3}\right)$.

f.   Using Wolfram|Alpha, the maximum is approximately 2.45 when $(x,y) = \left(\dfrac{10}{3}, \dfrac{10}{3}\right)$.

29.   The constraint equation is $x + y = 30$. Using Wolfram|Alpha, the maximum is approximately 980.833 when $(x,y) = \left(\dfrac{65}{9}, \dfrac{205}{9}\right) \approx (7,23)$. The maximum profit is approximately \$980,833 when 7 units of product A and 23 units of product B are produced.

31.   Finding the partial derivative: $f_x(x,y) = 42 \cdot \dfrac{3}{7} x^{-4/7} y^{4/7} = 18 x^{-4/7} y^{4/7} = \dfrac{18 y^{4/7}}{x^{4/7}}$

33.   Evaluating the expression: $f_x(x,y) dx = \dfrac{18(128)^{4/7}}{(2187)^{4/7}} \cdot (-4) = \dfrac{32}{9}(-4) = -\dfrac{128}{9} \approx -14.22$

7.5 The Total Differential

1.   First find the partial derivatives:
$$f_x(x,y) = 16x + 4 \qquad\qquad f_y(x,y) = -6y + 5$$
The total differential is therefore:
$$df = f_x(x,y)dx + f_y(x,y)dy = (16x+4)(1) + (-6y+5)(2) = 16x + 4 - 12y + 10 = 16x - 12y + 14$$

3.   First find the partial derivatives:
$$f_x(x,y) = 2x + 4ye^{xy} \qquad\qquad f_y(x,y) = -30y + 4xe^{xy}$$
Evaluating the partial derivatives:
$$f_x(3,1) = 2(3) + 4(1)e^{(3)(1)} = 6 + 4e^3 \qquad\qquad f_y(3,1) = -30(1) + 4(3)e^{(3)(1)} = -30 + 12e^3$$
The total differential is therefore:
$$df = f_x(3,1)dx + f_y(3,1)dy = (6 + 4e^3)(1) + (-30 + 12e^3)(1) = 6 + 4e^3 - 30 + 12e^3 = 16e^3 - 24 \approx 297.4$$

5. First find the partial derivatives:

$$f_x(x,y) = \frac{1}{2}x^{-1/2}y^{1/2} = \frac{\sqrt{y}}{2\sqrt{x}} \qquad\qquad f_y(x,y) = \frac{1}{2}x^{1/2}y^{-1/2} = \frac{\sqrt{x}}{2\sqrt{y}}$$

Evaluating the partial derivatives:

$$f_x(9,4) = \frac{\sqrt{4}}{2\sqrt{9}} = \frac{1}{3} \qquad\qquad f_y(9,4) = \frac{\sqrt{9}}{2\sqrt{4}} = \frac{3}{4}$$

The total differential is therefore:

$$df = f_x(9,4)dx + f_y(9,4)dy = \left(\frac{1}{3}\right)(1) + \left(\frac{3}{4}\right)(-1) = \frac{1}{3} - \frac{3}{4} = -\frac{5}{12} \approx -0.4167$$

7. First find the partial derivatives:

$$f_x(x,y,z) = 10xe^{2y-z^2} \qquad f_y(x,y,z) = 10x^2e^{2y-z^2} \qquad f_z(x,y,z) = -10x^2ze^{2y-z^2}$$

Evaluating the partial derivatives:

$$f_x(3,1,1) = 10(3)e^{2(1)-(1)^2} = 30e \qquad f_y(3,1,1) = 10(3)^2 e^{2(1)-(1)^2} = 90e \qquad f_z(3,1,1) = -10(3)^2(1)e^{2(1)-(1)^2} = -90e$$

The total differential is therefore:

$$\begin{aligned} df &= f_x(3,1,1)dx + f_y(3,1,1)dy + f_z(3,1,1)dz \\ &= (30e)(-2) + (90e)(1) + (-90e)(-1) \\ &= -60e + 90e + 90e \\ &= 120e \\ &\approx 326.19 \end{aligned}$$

9. First find the partial derivatives:

$$f_x(x,y) = -x^{-2}y = -\frac{y}{x^2} \qquad\qquad f_y(x,y) = x^{-1} = \frac{1}{x}$$

Evaluating the partial derivatives:

$$f_x(150,20) = -\frac{20}{(150)^2} = -\frac{1}{1125} \qquad\qquad f_y(150,20) = \frac{1}{150}$$

The total differential is therefore:

$$df = f_x(150,20)dx + f_y(150,20)dy = \left(-\frac{1}{1125}\right)(5) + \left(\frac{1}{150}\right)(1) = -\frac{1}{225} + \frac{1}{150} = \frac{1}{450} \approx 0.0022$$

11. First find the partial derivatives:

$$f_x(x,y,z) = \frac{y}{x} \qquad\qquad f_y(x,y,z) = \ln x \qquad\qquad f_z(x,y,z) = 2z$$

Evaluating the partial derivatives:

$$f_x(1,0,0) = \frac{0}{1} = 0 \qquad\qquad f_y(1,0,0) = \ln 1 = 0 \qquad\qquad f_z(1,0,0) = 2(0) = 0$$

The total differential is therefore:

$$df = f_x(1,0,0)dx + f_y(1,0,0)dy + f_z(1,0,0)dz = (0)(1) + (0)(2) + (0)(1) = 0 + 0 + 0 = 0$$

13. For $df(a,b) < 0$, we must have:

$$df(a,b) = f_x(a,b)dx + f_y(a,b)dy = f_x(a,b)(1) + f_y(a,b)(2) = f_x(a,b) + 2f_y(a,b)$$

Thus:

$$f_x(a,b) + 2f_y(a,b) < 0$$
$$2f_y(a,b) < -f_x(a,b)$$

15. For $df(a,b) < 0$, we must have:

$$df(a,b) = f_x(a,b)dx + f_y(a,b)dy = e^{a+b}dx + e^{a+b}dy = e^{a+b}(dx + dy)$$

So if $dx < 0$ and $dy < 0$, $df < 0$ since $e^{a+b} > 0$. It is not possible to have $df > 0$.

17. **a.** First find the partial derivatives:

$$f_x(x,y) = 60x^{-1/3}y^{1/3} = \frac{60y^{1/3}}{x^{1/3}} \qquad f_y(x,y) = 30x^{2/3}y^{-2/3} = \frac{30x^{2/3}}{y^{2/3}}$$

Evaluating the partial derivatives:

$$f_x(27,27) = \frac{60(27)^{1/3}}{(27)^{1/3}} = 60 \qquad f_y(27,27) = \frac{30(27)^{2/3}}{(27)^{2/3}} = 30$$

The total differential is therefore:

$$df = f_x(27,27)dx + f_y(27,27)dy = (60)(1) + (30)(-1) = 60 - 30 = 30$$

The approximate change in output is an increase of approximately 30 units.

b. The actual change is given by:

$$f(28,26) - f(27,27) = 90(28)^{2/3}(26)^{1/3} - 90(27)^{2/3}(27)^{1/3} \approx 28.51$$

The actual change in output is approximately 28.51 units, about 1.49 units less than the differential approximation.

19. **a.** First find the partial derivatives:

$$C_x(x,y) = 12x - 1 \qquad C_y(x,y) = 8y - 3$$

Evaluating the partial derivatives:

$$C_x(14,12) = 12(14) - 1 = 167 \qquad C_y(14,12) = 8(12) - 3 = 93$$

The total differential is therefore:

$$dC = C_x(14,12)dx + C_y(14,12)dy = (167)(2) + (93)(-2) = 334 - 186 = 148$$

The change in cost is approximately \$148.

b. The actual change is given by:

$$C(16,10) - C(14,12) = (6 \cdot 16^2 + 4 \cdot 10^2 - 16 - 3 \cdot 10) - (6 \cdot 14^2 + 4 \cdot 12^2 - 14 - 3 \cdot 12) = 1{,}890 - 1{,}702 = 188$$

The actual change in cost is \$188, about \$40 more than the differential approximation.

21. First find the partial derivatives:

$$f_x(x,y) = 0.8e^{0.03y} \qquad f_y(x,y) = 0.024xe^{0.03y}$$

Evaluating the partial derivatives:

$$f_x(200,40) = 0.8e^{0.03(40)} = 0.8e^{1.2} \qquad f_y(200,40) = 0.024(200)e^{0.03(40)} = 4.8e^{1.2}$$

The total differential is therefore:

$$df = f_x(200,40)dx + f_y(200,40)dy = (0.8e^{1.2})(-5) + (4.8e^{1.2})(0.5) = -4e^{1.2} + 2.4e^{1.2} \approx -5.31219$$

The approximate change in output is a decrease of 5,312.19 units.

23. First find the partial derivatives:

$$R_x(x,t) = [3.2x(-1.3) + 3.2(8.6 - 1.3x)]t^{1.2}e^{-1.8t} = 3.2(-2.6x + 8.6)t^{1.2}e^{-1.8t}$$

$$R_t(x,t) = 3.2x(8.6 - 1.3x)[-1.8t^{1.2}e^{-1.8t} + 1.2t^{0.2}e^{-1.8t}] = 3.2x(8.6 - 1.3x)(0.6t^{0.2}e^{-1.8t})(-3t + 2)$$

Evaluating the partial derivatives:

$$R_x(150,0.75) = 3.2(-2.6 \cdot 150 + 8.6)(0.75)^{1.2} e^{-1.8(0.75)} \approx -224.03$$

$$R_t(150,0.75) = 3.2(150)(8.6 - 1.3 \cdot 150)(0.6(0.75)^{0.2} e^{-1.8(0.75)})(-3(0.75) + 2) \approx 3{,}284.68$$

The total differential is therefore:

$$dR = R_x(150,0.75)dx + R_t(150,0.75)dt = (-224.03)(-5) + (3{,}284.68)(0.13) \approx 1{,}547.16$$

The approximate change in the person's reaction is an increase of 1,547.16 units.

25. First find the partial derivatives:

$$N_x(x,y) = 16.5(9.3y + 18.2)^{0.45}$$

$$N_y(x,y) = 7.425x(9.3y + 18.2)^{-0.55} \cdot 9.3 = 69.0525x(9.3y + 18.2)^{-0.55}$$

Evaluating the partial derivatives:

$$N_x(20,40) = 16.5(9.3 \cdot 40 + 18.2)^{0.45} \approx 241.8595$$

$$N_y(20,40) = 69.0525(20)(9.3 \cdot 40 + 18.2)^{-0.55} \approx 51.8802$$

The total differential is therefore:

$$dN = N_x(20,40)dx + N_y(20,40)dy = (241.8595)(-3) + (51.8802)(4) = -518.058$$

The approximate change in people reached is a decrease of 518,058 people.

27. Evaluating the integral: $\int_3^{10}(k + 4y)dy = ky + 2y^2\big|_3^{10} = (10k + 200) - (3k + 18) = 7k + 182$

29. Evaluating the integral:

$$\int_{x^3}^{4x}(3k^2 + 6ky^2)dy = (3k^2 y + 2ky^3)\big|_{x^3}^{4x}$$

$$= 3k^2(4x) + 2k(4x)^3 - 3k^2(x^3) - 2k(x^3)^3$$

$$= 12k^2 x + 128kx^3 - 3k^2 x^3 - 2kx^9$$

$$= kx(12k + 128x^2 - 3kx^2 - 2x^8)$$

31. Evaluating the integral: $\int_{9.999}^{10.001} 10,000\,dy = 10,000y\big|_{9.999}^{10.001} = 10,000(10.001 - 9.999) = 20$

33. Evaluating the integral: $-40(e^{-9} - 1)\int_0^{12}(e^{-0.5x})dx = \dfrac{-40(e^{-9} - 1)}{-0.5}e^{-0.5x}\bigg|_0^{12} = 80(e^{-9} - 1)(e^{-6} - 1) \approx 79.79$

7.6 Double Integrals as Volume

1. The volume is given by the integral: $\int_0^2 \int_0^1 7xy\,dy\,dx$

Finding the first integral: $\int_0^1 7xy\,dy = \dfrac{7}{2}xy^2\bigg|_0^1 = \dfrac{7}{2}x(1) - \dfrac{7}{2}x(0) = \dfrac{7}{2}x$

Now finding the double integral: $\int_0^2 \int_0^1 7xy\,dy\,dx = \int_0^2 \dfrac{7}{2}x\,dx = \dfrac{7}{4}x^2\bigg|_0^2 = \dfrac{7}{4}(4 - 0) = 7$

3. The volume is given by the integral: $\int_0^3 \int_0^2 (x - y)dy\,dx$

Finding the first integral: $\int_0^2 (x - y)dy = \left(xy - \dfrac{1}{2}y^2\right)\bigg|_0^2 = (2x - 2) - 0 = 2x - 2$

Now finding the double integral: $\int_0^3 \int_0^2 (x - y)dy\,dx = \int_0^3 (2x - 2)dx = (x^2 - 2x)\big|_0^3 = (9 - 6) - 0 = 3$

5. The volume is given by the integral: $\int_0^2 \int_0^1 (x + 3y)dy\,dx$

Finding the first integral: $\int_0^1 (x + 3y)dy = \left(xy + \dfrac{3}{2}y^2\right)\bigg|_0^1 = \left(x + \dfrac{3}{2}\right) - 0 = x + \dfrac{3}{2}$

Now finding the double integral: $\int_0^2 \int_0^1 (x + 3y)dy\,dx = \int_0^2 \left(x + \dfrac{3}{2}\right)dx = \left(\dfrac{1}{2}x^2 + \dfrac{3}{2}x\right)\bigg|_0^2 = (2 + 3) - 0 = 5$

7. The volume is given by the integral: $\int_0^4 \int_0^3 (6x^2 + y^2)\,dy\,dx$

Finding the first integral: $\int_0^3 (6x^2 + y^2)\,dy = \left(6x^2 y + \dfrac{1}{3}y^3\right)\Big|_0^3 = (18x^2 + 9) - 0 = 18x^2 + 9$

Now finding the double integral: $\int_0^4 \int_0^3 (6x^2 + y^2)\,dy\,dx = \int_0^4 (18x^2 + 9)\,dx = (6x^3 + 9x)\Big|_0^4 = (384 + 36) - 0 = 420$

9. The volume is given by the integral: $\int_0^1 \int_0^4 xy\,dy\,dx$

Finding the first integral: $\int_0^4 xy\,dy = \dfrac{1}{2}xy^2 \Big|_0^4 = 8x - 0 = 8x$

Now finding the double integral: $\int_0^1 \int_0^4 xy\,dy\,dx = \int_0^1 8x\,dx = 4x^2 \Big|_0^1 = 4 - 0 = 4$

11. The volume is given by the integral: $\int_0^2 \int_x^3 4ye^y\,dy\,dx$

Finding the first integral:

$\int_x^3 4ye^y\,dy = 4ye^y \Big|_x^3 - \int_x^3 4e^y\,dy = \left(4ye^y - 4e^y\right)\Big|_x^3 = (12e^3 - 4e^3) - (4xe^x - 4e^x) = 8e^3 - 4xe^x + 4e^x$

Now finding the double integral:

$$\int_0^2 \int_x^3 4ye^y\,dy\,dx = \int_0^2 (8e^3 - 4xe^x + 4e^x)\,dx$$
$$= \left(8e^3 x - 4(xe^x - e^x) + 4e^x\right)\Big|_0^2$$
$$= \left(8e^3 x - 4xe^x + 8e^x\right)\Big|_0^2$$
$$= (16e^3 - 8e^2 + 8e^2) - 8$$
$$= 16e^3 - 8$$
$$\approx 313.369$$

13. The volume is given by the integral: $\int_0^5 \int_0^{\sqrt{y}} \dfrac{4x}{y}\,dx\,dy$

Finding the first integral: $\int_0^{\sqrt{y}} \dfrac{4x}{y}\,dx = \dfrac{2x^2}{y}\Big|_0^{\sqrt{y}} = \dfrac{2y}{y} - 0 = 2$

Now finding the double integral: $\int_0^5 \int_0^{\sqrt{y}} \dfrac{4x}{y}\,dx\,dy = \int_0^5 2\,dy = 2y\Big|_0^5 = 10 - 0 = 10$

15. The volume is given by the integral: $\int_0^1 \int_0^1 e^{x+y}\,dx\,dy$

Finding the first integral: $\int_0^1 e^{x+y}\,dx = e^{x+y}\Big|_0^1 = e^{1+y} - e^y$

Now finding the double integral: $\int_0^1 \int_0^1 e^{x+y}\,dx\,dy = \int_0^1 (e^{1+y} - e^y)\,dy = (e^{1+y} - e^y)\Big|_0^1 = (e^2 - e) - (e - 1) = e^2 - 2e + 1 \approx 2.952$

17. The volume is given by the integral: $\int_0^4 \int_0^{20} \frac{20,000e^y}{1+x} dx\, dy$

Finding the first integral: $\int_0^{20} \frac{20,000e^y}{1+x} dx = 20,000e^y \ln(1+x)\Big|_0^{20} = (20,000\ln 21)e^y$

Now finding the double integral:

$$\int_0^4 \int_0^{20} \frac{20,000e^y}{1+x} dx\, dy = \int_0^4 (20,000\ln 21)e^y\, dy$$

$$= (20,000\ln 21)e^y\Big|_0^4$$

$$= (20,000\ln 21)(e^4 - 1)$$

$$\approx 3,263,615.41$$

19. Evaluating the integral: $\int (12xy^3 + 8xy) dx = 12y^3 \cdot \frac{1}{2}x^2 + 8y \cdot \frac{1}{2}x^2 + g(y) = 6x^2y^3 + 4x^2y + g(y)$

21. Evaluating the integral: $\int 4xe^x\, dy = 4xe^x \cdot y + g(x) = 4xye^x + g(x)$

23. Finding the first integral: $\int_0^1 xy\, dy = \frac{1}{2}xy^2\Big|_0^1 = \frac{1}{2}x(1-0) = \frac{1}{2}x$

Now finding the double integral: $\int_1^2 \int_0^1 xy\, dy\, dx = \int_1^2 \frac{1}{2}x\, dx = \frac{1}{4}x^2\Big|_1^2 = 1 - \frac{1}{4} = \frac{3}{4}$

25. Finding the first integral: $\int_1^3 dx = x\Big|_1^3 = 3 - 1 = 2$

Now finding the double integral: $\int_0^2 \int_1^3 dx\, dy = \int_0^2 2\, dy = 2y\Big|_0^2 = 4 - 0 = 4$

27. Finding the first integral: $\int_0^1 (y-x) dx = \left(xy - \frac{1}{2}x^2\right)\Big|_0^1 = y - \frac{1}{2}$

Now finding the double integral: $\int_1^2 \int_0^1 (y-x) dx\, dy = \int_1^2 \left(y - \frac{1}{2}\right) dy = \left(\frac{1}{2}y^2 - \frac{1}{2}y\right)\Big|_1^2 = (2-1) - \left(\frac{1}{2} - \frac{1}{2}\right) = 1$

29. Finding the first integral: $\int_2^3 \frac{2y}{x} dy = \frac{y^2}{x}\Big|_2^3 = \frac{9}{x} - \frac{4}{x} = \frac{5}{x}$

Now finding the double integral: $\int_1^2 \int_2^3 \frac{2y}{x} dy\, dx = \int_1^2 \frac{5}{x} dx = 5\ln x\Big|_1^2 = 5\ln 2 \approx 3.466$

31. Finding the first integral: $\int_0^y (x + 3y) dx = \left(\frac{1}{2}x^2 + 3xy\right)\Big|_0^y = \frac{1}{2}y^2 + 3y^2 = \frac{7}{2}y^2$

Now finding the double integral: $\int_1^2 \int_0^y (x + 3y) dx\, dy = \int_1^2 \frac{7}{2}y^2\, dy = \frac{7}{6}y^3\Big|_1^2 = \frac{28}{3} - \frac{7}{6} = \frac{49}{6} \approx 8.167$

33. Finding the first integral: $\int_0^x e^{x^2} dy = e^{x^2}y\Big|_0^x = xe^{x^2}$

Now finding the double integral: $\int_0^1 \int_0^x e^{x^2} dy\, dx = \int_0^1 xe^{x^2} dx = \frac{1}{2}e^{x^2}\Big|_0^1 = \frac{1}{2}(e-1) \approx 0.859$

35. Finding the first integral: $\int_0^{x/2} e^{2y-x} dy = \frac{1}{2}e^{2y-x}\Big|_0^{x/2} = \frac{1}{2}(1 - e^{-x})$

Now finding the double integral: $\int_0^1 \int_0^{x/2} e^{2y-x} dy\, dx = \int_0^1 \frac{1}{2}(1 - e^{-x}) dx = \frac{1}{2}(x + e^{-x})\Big|_0^1 = \frac{1}{2}(1 + e^{-1}) - \frac{1}{2} = \frac{1}{2e} \approx 0.184$

37. Finding the first integral: $\displaystyle\int_0^{x+2} \frac{y}{x+2}\,dy = \frac{y^2}{2(x+2)}\bigg|_0^{x+2} = \frac{x+2}{2}$

Now finding the double integral: $\displaystyle\int_0^3\int_0^{x+2} \frac{y}{x+2}\,dy\,dx = \int_0^3 \frac{x+2}{2}\,dx = \frac{(x+2)^2}{4}\bigg|_0^3 = \frac{25}{4} - 1 = \frac{21}{4}$

39. Finding the first integral: $\displaystyle\int_0^{\sqrt{1-x^2}} (2x+y)\,dy = \left(2xy + \frac{1}{2}y^2\right)\bigg|_0^{\sqrt{1-x^2}} = 2x\left(1-x^2\right)^{1/2} + \frac{1}{2}\left(1-x^2\right)$

Now finding the double integral:

$$\int_0^1\int_0^{\sqrt{1-x^2}} (2x+y)\,dy\,dx = \int_0^1 \left(2x\left(1-x^2\right)^{1/2} + \frac{1}{2}\left(1-x^2\right)\right)dx = -\frac{\left(1-x^2\right)^{3/2}}{3/2} + \frac{x}{2} - \frac{x^3}{6}\bigg|_0^1 = \left(0 + \frac{1}{2} - \frac{1}{6}\right) - \left(-\frac{2}{3}\right) = 1$$

41. We use the uniform probability density model with the following probability density functions:

$P(X) = \dfrac{1}{2.1-1.9} = \dfrac{1}{0.2} = 5$ $\qquad\qquad$ $P(Y) = \dfrac{1}{1.1-0.9} = \dfrac{1}{0.2} = 5$

$P(X \text{ and } Y) = P(X)P(Y) = 5\cdot 5 = 25$

The probability is given by the integral: $\displaystyle\int_{1.95}^{2.05}\int_{0.95}^{1.05} 25\,dy\,dx$

Finding the first integral: $\displaystyle\int_{0.95}^{1.05} 25\,dy = 25y\big|_{0.95}^{1.05} = 26.25 - 23.75 = 2.5$

Now finding the double integral: $\displaystyle\int_{1.95}^{2.05}\int_{0.95}^{1.05} 25\,dy\,dx = \int_{1.95}^{2.05} 2.5\,dx = 2.5x\big|_{1.95}^{2.05} = 5.125 - 4.875 = 0.25 = \frac{1}{4}$

43. We use the uniform probability density model with the following probability density functions:

$P(X) = \dfrac{1}{11-9} = \dfrac{1}{2} = 0.5$ $\qquad\qquad$ $P(Y) = \dfrac{1}{0.210-0.190} = \dfrac{1}{0.02} = 50$

$P(X \text{ and } Y) = P(X)P(Y) = 0.5\cdot 50 = 25$

The probability is given by the integral: $\displaystyle\int_{9.5}^{10.5}\int_{0.195}^{0.205} 25\,dy\,dx$

Finding the first integral: $\displaystyle\int_{0.195}^{0.205} 25\,dy = 25y\big|_{0.195}^{0.205} = 5.125 - 4.875 = 0.25$

Now finding the double integral: $\displaystyle\int_{9.5}^{10.5}\int_{0.195}^{0.205} 25\,dy\,dx = \int_{9.5}^{10.5} 0.25\,dx = 0.25x\big|_{9.5}^{10.5} = 2.625 - 2.375 = 0.25 = \frac{1}{4}$

45. Evaluating the integral: $\displaystyle\int\left(3x^2 + 6xy^2\right)dy = 3x^2 y + 6x\cdot\frac{1}{3}y^3 + g(y) = 3x^2 y + 2xy^3 + g(x)$

47. The volume is given by the integral: $\displaystyle\int_0^3\int_0^2 900 x^{0.85} y^{0.15}\,dx\,dy$

Finding the first integral: $\displaystyle\int_0^2 900 x^{0.85} y^{0.15}\,dx = \frac{900 x^{1.85} y^{0.15}}{1.85}\bigg|_0^2 = \frac{900(2)^{1.85}}{1.85} y^{0.15}$

Now finding the double integral:

$$\int_0^3\int_0^2 900 x^{0.85} y^{0.15}\,dx\,dy = \int_0^3 \frac{900(2)^{1.85}}{1.85} y^{0.15}\,dy = \frac{900(2)^{1.85} y^{1.15}}{(1.85)(1.15)}\bigg|_0^3 = \frac{900(2)^{1.85}(3)^{1.15}}{(1.85)(1.15)} \approx 5{,}394.71$$

49. The probability is given by the integral: $\int_0^5 \int_0^3 300 e^{-0.2x-0.9y}\, dy\, dx$

Finding the first integral: $\int_0^3 300 e^{-0.2x-0.9y}\, dy = -\dfrac{1,000}{3} e^{-0.2x-0.9y}\Big|_0^3 = -\dfrac{1,000}{3}\left(e^{-0.2x-2.7} - e^{-0.2x}\right)$

Now finding the double integral:

$$\int_0^5 \int_0^3 300 e^{-0.2x-0.9y}\, dy\, dx = \int_0^5 -\dfrac{1,000}{3}\left(e^{-0.2x-2.7} - e^{-0.2x}\right) dx$$

$$= \dfrac{5,000}{3}\left(e^{-0.2x-2.7} - e^{-0.2x}\right)\Big|_0^5$$

$$= \dfrac{5,000}{3}\left(e^{-3.7} - e^{-1}\right) - \dfrac{5,000}{3}\left(e^{-2.7} - 1\right)$$

$$\approx 982.731$$

The mold population is approximately 982.731 million spores.

51. The average is: $\dfrac{2+4+6+8+10}{5} = \dfrac{30}{5} = 6$

53. The heights of the rectangles are 14, 28, 42, 56, and 70. The average is: $\dfrac{14+28+42+56+70}{5} = \dfrac{210}{5} = 42$

55. Evaluating the integral: $\dfrac{1}{6-3}\int_3^6 3x^2\, dx = \int_3^6 x^2\, dx = \dfrac{1}{3}x^3\Big|_3^6 = \dfrac{1}{3}\left(6^3 - 3^3\right) = 63$

57. Evaluating the integral: $\dfrac{1}{7-3}\int_3^7 1,000 e^{-0.50x}\, dx = 250\int_3^7 e^{-0.50x}\, dx = -500 e^{-0.50x}\Big|_3^7 = -500\left(e^{-3.5} - e^{-1.5}\right) \approx 96.466$

59. Evaluating the expression: $\dfrac{f(0)+f(1)+f(2)+f(3)}{4} = \dfrac{5+6+7+8}{4} = \dfrac{26}{4} = 6.5$

7.7 The Average Value of a Function

1. Finding the average value: $\dfrac{1}{4-0}\int_0^4 \left(x^2+1\right) dx = \dfrac{1}{4}\int_0^4 \left(x^2+1\right) dx = \dfrac{1}{4}\left(\dfrac{1}{3}x^3 + x\right)\Big|_0^4 = \dfrac{1}{4}\left(\dfrac{64}{3} + 4\right) = \dfrac{19}{3}$

3. Finding the average value: $\dfrac{1}{16-0}\int_0^{16} \sqrt{x}\, dx = \dfrac{1}{16}\int_0^{16} x^{1/2}\, dx = \dfrac{1}{16}\cdot\dfrac{2}{3}x^{3/2}\Big|_0^{16} = \dfrac{1}{24}\left(16^{3/2} - 0\right) = \dfrac{8}{3}$

5. Finding the average value: $\dfrac{1}{50-20}\int_{20}^{50} 440 e^{-0.04x}\, dx = \dfrac{44}{3}\int_{20}^{50} e^{-0.04x}\, dx = \dfrac{44}{3}\cdot\dfrac{e^{-0.04x}}{-0.04}\Big|_{20}^{50} = -\dfrac{1,100}{3}\left(e^{-2} - e^{-0.8}\right) \approx 115.131$

7. Finding the average value: $\dfrac{1}{5-1}\int_1^5 x\sqrt{x^2-1}\, dx = \dfrac{1}{4}\int_1^5 x\left(x^2-1\right)^{1/2} dx = \dfrac{1}{8}\cdot\dfrac{2}{3}\left(x^2-1\right)^{3/2}\Big|_1^5 = \dfrac{1}{12}\left(24^{3/2} - 0\right) \approx 9.798$

9. This integral gives us the average pollution units per mile between 3 and 10 miles from the facility.

11. This integral gives us the average depreciation per month between 12 and 24 months after the equipment was purchased.

13. This integral gives us the average value of the piece of art during the first 20 years after it is purchased.

15. This integral gives us the average production per day for an employee who has been on the job between 60 and 120 days.

17. Finding the average value:

$$\dfrac{1}{4-0}\int_0^4 230,000 e^{-0.02t}\, dt = 57,500\int_0^4 e^{-0.02t}\, dt = -2,875,000 e^{-0.02t}\Big|_0^4 = -2,875,000\left(e^{-0.08} - 1\right) \approx 221,040$$

The average value of a house is $221,040 during the first four years of the recession.

19. Finding the average value:

$$\frac{1}{500-400}\int_{400}^{500}(-0.06x+35)\,dx = \frac{1}{100}\int_{400}^{500}(-0.06x+35)\,dx = \frac{1}{100}\left(-0.03x^2+35x\right)\Big|_{400}^{500} = \frac{1}{100}(10{,}000-9{,}200) = 8$$

The average marginal cost is \$8 per unit for units 400 through 500 of production.

21. The average value is the double integral: $\dfrac{1}{(12-8)(3-1)}\int_8^{12}\int_1^3 800x^{2/3}y^{1/3}\,dy\,dx = 100\int_8^{12}\int_1^3 x^{2/3}y^{1/3}\,dy\,dx$

Finding the first integral: $\displaystyle\int_1^3 x^{2/3}y^{1/3}\,dy = \frac{3}{4}x^{2/3}y^{4/3}\Big|_1^3 = \frac{3}{4}x^{2/3}\left(3^{4/3}-1\right) = \frac{3\left(3^{4/3}-1\right)}{4}x^{2/3}$

Now finding the double integral:

$$100\int_8^{12}\int_1^3 x^{2/3}y^{1/3}\,dy\,dx = 100\int_8^{12}\frac{3\left(3^{4/3}-1\right)}{4}x^{2/3}\,dx$$

$$= 75\left(3^{4/3}-1\right)\cdot\frac{3}{5}x^{5/3}\Big|_8^{12}$$

$$= 45\left(3^{4/3}-1\right)\left(12^{5/3}-8^{5/3}\right)$$

$$\approx 4{,}625.51$$

The average number of units produced is 4,625.5 units.

23. First simplify the revenue function:

$$R(x_A,x_B) = x_A p_A + x_B p_B = (430-4p_A)p_A + (560-3p_B)p_B = 430p_A - 4p_A^2 + 560p_B - 3p_B^2$$

The average value is the double integral:

$$\frac{1}{(110-80)(50-40)}\int_{80}^{110}\int_{40}^{50}\left(430p_A - 4p_A^2 + 560p_B - 3p_B^2\right)dp_B\,dp_A = \frac{1}{300}\int_{80}^{110}\int_{40}^{50}\left(430p_A - 4p_A^2 + 560p_B - 3p_B^2\right)dp_B\,dp_A$$

Finding the first integral:

$$\int_{40}^{50}\left(430p_A - 4p_A^2 + 560p_B - 3p_B^2\right)dp_B$$

$$= \left(430p_A p_B - 4p_A^2 p_B + 280p_B^2 - p_B^3\right)\Big|_{40}^{50}$$

$$= \left(21{,}500p_A - 200p_A^2 + 700{,}000 - 125{,}000\right) - \left(17{,}200p_A - 160p_A^2 + 448{,}000 - 64{,}000\right)$$

$$= 4{,}300p_A - 40p_A^2 + 191{,}000$$

Now finding the double integral:

$$\frac{1}{300}\int_{80}^{110}\int_{40}^{50}\left(430p_A - 4p_A^2 + 560p_B - 3p_B^2\right)dp_B\,dp_A = \frac{1}{300}\int_{80}^{110}\left(4{,}300p_A - 40p_A^2 + 191{,}000\right)dp_A$$

$$= \frac{1}{300}\left(2{,}150p_A^2 - \frac{40}{3}p_A^3 + 191{,}000p_A\right)\Big|_{80}^{110}$$

$$= \frac{1}{300}(29{,}278{,}333.33 - 22{,}213{,}333.33)$$

$$= 23{,}550$$

The average revenue is \$23,550 per month.

25. The average value is the double integral:

$$\frac{1}{(5-4)(20-16)}\int_4^5\int_{16}^{20}\left(0.04x^2+0.08xy+0.006y^2\right)dy\,dx = \int_4^5\int_{16}^{20}\left(0.01x^2+0.02xy+0.0015y^2\right)dy\,dx$$

Finding the first integral:

$$\int_{16}^{20}\left(0.01x^2+0.02xy+0.0015y^2\right)dy = \left(0.01x^2y+0.01xy^2+0.0005y^3\right)\Big|_{16}^{20}$$

$$= \left(0.2x^2+4x+4\right)-\left(0.16x^2+2.56x+2.048\right)$$

$$= 0.04x^2+1.44x+1.952$$

Now finding the double integral:

$$\int_4^5\int_{16}^{20}\left(0.01x^2+0.02xy+0.0015y^2\right)dy\,dx = \int_4^5\left(0.04x^2+1.44x+1.952\right)dx$$

$$= \left(\frac{0.04}{3}x^3+0.72x^2+1.952x\right)\Big|_4^5$$

$$= \left(\frac{5}{3}+18+9.76\right)-\left(\frac{2.56}{3}+11.52+7.808\right)$$

$$\approx 9.25$$

The average pollution is 9.25 units per day.

27. The average value is the double integral:

$$\frac{1}{(70-50)(350-200)}\int_{50}^{70}\int_{200}^{350}\left(120x+270y-x^2-y^2\right)dy\,dx = \frac{1}{3,000}\int_{50}^{70}\int_{200}^{350}\left(120x+270y-x^2-y^2\right)dy\,dx$$

Finding the first integral:

$$\int_{200}^{350}\left(120x+270y-x^2-y^2\right)dy$$

$$= \left(120xy+135y^2-x^2y-\frac{1}{3}y^3\right)\Big|_{200}^{350}$$

$$= \left(42,000x+16,537,500-350x^2-\frac{42,875,000}{3}\right)-\left(24,000x+5,400,000-200x^2-\frac{8,000,000}{3}\right)$$

$$= 18,000x-150x^2-487,500$$

Now finding the double integral:

$$\frac{1}{3,000}\int_{50}^{70}\int_{200}^{350}\left(120x+270y-x^2-y^2\right)dy\,dx = \frac{1}{3,000}\int_{50}^{70}\left(18,000x-150x^2-487,500\right)dx$$

$$= \int_{50}^{70}\left(6x-\frac{1}{20}x^2-162.5\right)dx$$

$$= \left(3x^2-\frac{1}{60}x^3-162.5x\right)\Big|_{50}^{70}$$

$$= \left(14,700-\frac{34,300}{6}-11,375\right)-\left(7,500-\frac{12,500}{6}-8,125\right)$$

$$\approx 316.67$$

The average weekly revenue is $31,667.

29. The average value is the double integral:

$$\frac{1}{(20-12)(24-20)}\int_{12}^{20}\int_{20}^{24}20x(4y+20)\,dy\,dx = \frac{5}{2}\int_{12}^{20}\int_{20}^{24}x(y+5)\,dy\,dx$$

Finding the first integral:

$$\int_{20}^{24}x(y+5)\,dy = x\left(\frac{1}{2}y^2 + 5y\right)\Big|_{20}^{24} = x(288+120) - x(200+100) = 108x$$

Now finding the double integral:

$$\frac{5}{2}\int_{12}^{20}\int_{20}^{24}x(y+5)\,dy\,dx = \frac{5}{2}\int_{12}^{20}108x\,dx$$

$$= 270\int_{12}^{20}x\,dx$$

$$= 135x^2\Big|_{12}^{20}$$

$$= 54,000 - 19,440$$

$$= 34,560$$

The average number of people reached by the advertising is 34,560 people.

31. The average value is the double integral:

$$\frac{1}{(2-1.5)(38-30)}\int_{1.5}^{2}\int_{30}^{38}0.6xe^{0.04y}\,dy\,dx = 0.15\int_{1.5}^{2}\int_{30}^{38}xe^{0.04y}\,dy\,dx$$

Finding the first integral:

$$\int_{30}^{38}xe^{0.04y}\,dy = \left(25xe^{0.04y}\right)\Big|_{30}^{38} = 25xe^{1.52} - 25xe^{1.2} = 25\left(e^{1.52} - e^{1.2}\right)x$$

Now finding the double integral:

$$0.15\int_{1.5}^{2}\int_{30}^{38}xe^{0.04y}\,dy\,dx = 0.15\int_{1.5}^{2}25\left(e^{1.52} - e^{1.2}\right)x\,dx$$

$$= 3.75\left(e^{1.52} - e^{1.2}\right)\int_{1.5}^{2}x\,dx$$

$$= 1.875\left(e^{1.52} - e^{1.2}\right)x^2\Big|_{1.5}^{2}$$

$$= 1.875\left(e^{1.52} - e^{1.2}\right)(4 - 2.25)$$

$$\approx 4.108$$

The average weekly output is 4.108 units.

33. Using Wolfram/Alpha to evaluate the average values:

a. $\dfrac{1}{12-0}\int_{0}^{12}2.5te^{-0.07t}\,dt = \dfrac{1}{12}\int_{0}^{12}2.5te^{-0.07t}\,dt \approx 8.7437$ œg/mL

b. $\dfrac{1}{24-12}\int_{12}^{24}2.5te^{-0.07t}\,dt = \dfrac{1}{12}\int_{12}^{24}2.5te^{-0.07t}\,dt \approx 12.5368$ œg/mL

c. $\dfrac{1}{24-0}\int_{0}^{24}2.5te^{-0.07t}\,dt = \dfrac{1}{24}\int_{0}^{24}2.5te^{-0.07t}\,dt \approx 10.6403$ œg/mL

Chapter 7 Test

1. The function $R(x)$ represents the revenue realized as a function of newspaper advertisements only, while $R(x,y)$ represents the revenue realized as a function of both newspaper and radio advertising.

2. Finding the value: $f(0,2) = 4(0)e^{0+2} = 0$

3. The partial derivative $\dfrac{\partial}{\partial x}N(4,10) = 15$ means that as x changes from 4 to 5 (while y remains constant at 10), $N(x,y)$ will increase by about 15. The partial derivative $\dfrac{\partial}{\partial y}N(4,10) = -4$ means that as y changes from 10 to 11 (while x remains constant at 4), $N(x,y)$ will decrease by about 4.

4. We would expect $D_x(x,y) < 0$ since the demand should decrease as the price increases. We would expect $D_y(x,y) > 0$ since the demand should increase as the competitor's price increases.

5. a. Finding the partial derivative: $\dfrac{\partial}{\partial w}\text{BMI}(w,h) = \dfrac{703}{h^2} > 0$

   The positive sign tells us as the weight of a person increases and height remains constant, the body mass index should increase.

   b. Finding the partial derivative: $\dfrac{\partial}{\partial h}\text{BMI}(w,h) = \dfrac{-1,406w}{h^3} < 0$

   The negative sign tells us as the height of a person increases and weight remains constant, the body mass index should decrease.

6. Finding the partial derivatives:

   $f_x(x,y) = 12x + 5y$ $\qquad\qquad\qquad$ $f_y(x,y) = -12y + 5x$

   Evaluating the partial derivatives:

   $f_x(2,1) = 12(2) + 5(1) = 24 + 5 = 29$ $\qquad$ $f_y(2,1) = -12(1) + 5(2) = -12 + 10 = -2$

7. Finding the partial derivatives:

   $f_x(x,y) = 2x + 3$ $\qquad\qquad\qquad$ $f_y(x,y) = 14y + 4$

   Setting the partial derivatives equal to 0:

   $2x + 3 = 0$ $\qquad\qquad\qquad\qquad$ $14y + 4 = 0$

   $2x = -3$ $\qquad\qquad\qquad\qquad\quad$ $14y = -4$

   $x = -\dfrac{3}{2}$ $\qquad\qquad\qquad\qquad\quad$ $y = -\dfrac{2}{7}$

   The partial derivatives are equal to 0 at the point $\left(-\dfrac{3}{2}, -\dfrac{2}{7}\right)$.

8. a. Finding the partial derivative: $f_x(x,y) = 20x^{-2/3}y^{2/3}$

   Evaluating the partial derivative: $f_x(64,27) = 20(64)^{-2/3}(27)^{2/3} = \dfrac{45}{4} = 11.25$

   The production will increase by approximately 11.25 units.

   b. Finding the partial derivative: $f_y(x,y) = 40x^{1/3}y^{-1/3}$

   Evaluating the partial derivative: $f_y(64,27) = 40(64)^{1/3}(27)^{-1/3} = \dfrac{160}{3} \approx 53.33$

   The production will decrease by approximately 53.33 units.

9. Evaluating the expression:

   $D(x,y) = f_{xx}(x,y) \cdot f_{yy}(x,y) - f_{xy}^2(x,y) = (7)(-6) - (0)^2 = -42$

   Since $D(x,y) = -42 < 0$, $\left(-\dfrac{4}{7}, \dfrac{5}{6}\right)$ is a saddle point.

10. Finding the first and second derivatives:

$$f_x(x,y) = 6x - 12 \qquad f_y(x,y) = -2y + 16$$

$$f_{xx}(x,y) = 6 \qquad f_{yy}(x,y) = -2 \qquad f_{xy}(x,y) = 0$$

Finding the critical points by setting the first derivatives equal to 0 results in the equations:

$$6x - 12 = 0 \qquad\qquad -2y + 16 = 0$$
$$6x = 12 \qquad\qquad -2y = -16$$
$$x = 2 \qquad\qquad y = 8$$

Now evaluating the expression:

$$D(x,y) = f_{xx}(x,y) \cdot f_{yy}(x,y) - f_{xy}^2(x,y) = (6)(-2) - (0)^2 = -12$$

Since $D(2,8) = -12 < 0$, $(2,8)$ is a saddle point.

11. Finding the first and second derivatives:

$$f_x(x,y) = 4x^3 - 4x \qquad f_y(x,y) = 2y$$

$$f_{xx}(x,y) = 12x^2 - 4 \qquad f_{yy}(x,y) = 2 \qquad f_{xy}(x,y) = 0$$

Finding the critical points by setting the first derivatives equal to 0 results in the equations:

$$4x^3 - 4x = 0 \qquad\qquad 2y = 0$$
$$4x(x+1)(x-1) = 0 \qquad\qquad y = 0$$
$$x = -1, 0, 1$$

There are three critical points, at $(-1,0)$, $(0,0)$, and $(1,0)$.
Now finding the expression:

$$D(x,y) = f_{xx}(x,y) \cdot f_{yy}(x,y) - f_{xy}^2(x,y) = (12x^2 - 4)(2) - (0)^2 = 24x^2 - 8$$

Evaluating at each critical point:

$$D(-1,0) = 24(-1)^2 - 8 = 16 > 0$$
$$D(0,0) = 24(0)^2 - 8 = -8 < 0$$
$$D(1,0) = 24(1)^2 - 8 = 16 > 0$$

Since $D(-1,0) > 0$ and $f_{xx}(-1,0) = 8 > 0$, $(-1,0)$ is a local minimum point.

Since $D(0,0) < 0$, $(0,0)$ is a saddle point.

Since $D(1,0) > 0$ and $f_{xx}(1,0) = 8 > 0$, $(1,0)$ is a local minimum point.

12. If the amount of dollars allocated to labor and capital is increased by $1, from \$531,000 to \$531,001, then the number of units produced would increase by approximately 4.3.

13. Construct the Lagrangian function:

$$F(x,y,\lambda) = f(x,y) + \lambda g(x,y) = 300x^{1/3}y^{2/3} + \lambda(x + 2y - 300)$$

Finding the first-order partial derivatives and setting them equal to 0:

$$F_x = 100x^{-2/3}y^{2/3} + \lambda = 0$$
$$F_y = 200x^{1/3}y^{-1/3} + 2\lambda = 0$$
$$F_\lambda = x + 2y - 300 = 0$$

Solving for λ in the first two equations yields:

$$\lambda = -\frac{100y^{2/3}}{x^{2/3}}$$

$$\lambda = -\frac{100x^{1/3}}{y^{1/3}}$$

Therefore:

$$-\frac{100y^{2/3}}{x^{2/3}} = -\frac{100x^{1/3}}{y^{1/3}}$$

$$y = x$$

Substituting into $x + 2y = 300$:

$$x + 2x = 300$$
$$3x = 300$$
$$x = 100$$
$$y = 100$$

Evaluating the function: $f(100,100) = 300(100)^{1/3}(100)^{2/3} = 30,000$

14. **a.** The constraint equation is $75x + 125y = 60,000$, which simplifies to $3x + 5y = 2,400$.

Construct the Lagrangian function:

$$F(x,y,\lambda) = f(x,y) + \lambda g(x,y) = 40x^{1/4}y^{3/4} + \lambda(3x + 5y - 2,400)$$

Finding the first-order partial derivatives and setting them equal to 0:

$$F_x = 10x^{-3/4}y^{3/4} + 3\lambda = 0$$
$$F_y = 30x^{1/4}y^{-1/4} + 5\lambda = 0$$
$$F_\lambda = 3x + 5y - 2,400 = 0$$

Solving for λ in the first two equations yields:

$$\lambda = -\frac{10y^{3/4}}{3x^{3/4}}$$

$$\lambda = -\frac{30x^{1/4}}{5y^{1/4}}$$

Therefore:

$$-\frac{10y^{3/4}}{3x^{3/4}} = -\frac{30x^{1/4}}{5y^{1/4}}$$

$$50y = 90x$$

$$y = \frac{9}{5}x$$

Substituting into $3x + 5y = 2,400$:

$$3x + 5\left(\frac{9}{5}x\right) = 2,400$$

$$12x = 2,400$$

$$x = 200$$

$$y = \frac{9}{5}(200) = 360$$

They should allocate 200 units to labor ($15,000) and 360 units to capital ($45,000).

b. Evaluating the function: $f(200,360) = 40(200)^{1/4}(360)^{3/4} \approx 12,432$ units

15. First find the partial derivatives:

$$f_x(x,y) = 10x + 11 \qquad\qquad f_y(x,y) = -8y - 6$$

Evaluating the partial derivatives:

$$f_x(3,8) = 10(3) + 11 = 30 + 11 = 41 \qquad f_y(3,8) = -8(8) - 6 = -64 - 6 = -70$$

The total differential is therefore:

$$df = f_x(3,8)dx + f_y(3,8)dy = (41)(0.2) + (-70)(-0.1) = 8.2 + 7.0 = 15.2$$

16.　**a.**　First find the partial derivatives:

$$f_x(x,y) = 160x^{-3/5}y^{3/5} = \frac{160y^{3/5}}{x^{3/5}} \qquad\qquad f_y(x,y) = 240x^{2/5}y^{-2/5} = \frac{240x^{2/5}}{y^{2/5}}$$

Evaluating the partial derivatives:

$$f_x(11,40) = \frac{160(40)^{3/5}}{(11)^{3/5}} \approx 347.153 \qquad\qquad f_y(11,40) = \frac{240(11)^{2/5}}{(40)^{2/5}} \approx 143.200$$

The total differential is therefore:

$$df = f_x(11,40)dx + f_y(11,40)dy = (347.153)(-1) + (143.200)(2) = -347.153 + 286.400 \approx -60.75$$

The approximate change in output is a decrease of approximately 60.75 units.

b.　The actual change is given by:

$$f(10,42) - f(11,40) = 400(10)^{2/5}(42)^{3/5} - 400(11)^{2/5}(40)^{3/5} \approx -84.12$$

The actual decrease in output is approximately 84.12 units, about 23.37 units more than the differential approximation.

17.　Evaluating the integral: $\int 16x^3y^2\,dx = 16y^2 \cdot \frac{1}{4}x^4 + g(y) = 4x^4y^2 + g(y)$

18.　This sum represents the integral $\iint f(x,y)\,dy\,dx$.

19.　Finding the first integral: $\int_1^2 x^2y\,dx = \left(\frac{1}{3}x^3y\right)\Big|_1^2 = \frac{8}{3}y - \frac{1}{3}y = \frac{7}{3}y$

Now finding the double integral: $\int_1^5 \int_1^2 x^2y\,dx\,dy = \int_1^5 \frac{7}{3}y\,dy = \frac{7}{6}y^2\Big|_1^5 = \frac{175}{6} - \frac{7}{6} = \frac{168}{6} = 28$

20.　Finding the first integral: $\int_0^{\ln x} \frac{y}{x}\,dy = \frac{y^2}{2x}\Big|_0^{\ln x} = \frac{(\ln x)^2}{2x}$

Now finding the double integral: $\int_1^e \int_0^{\ln x} \frac{y}{x}\,dy\,dx = \int_1^e \frac{(\ln x)^2}{2x}\,dy = \frac{(\ln x)^3}{6}\Big|_1^e = \frac{1}{6} - 0 = \frac{1}{6}$

21.　The volume is given by the integral: $\int_0^1 \int_1^2 (x^2 + xy^3)\,dx\,dy$

Finding the first integral: $\int_1^2 (x^2 + xy^3)\,dx = \left(\frac{1}{3}x^3 + \frac{1}{2}x^2y^3\right)\Big|_1^2 = \left(\frac{8}{3} + 2y^3\right) - \left(\frac{1}{3} + \frac{1}{2}y^3\right) = \frac{7}{3} + \frac{3}{2}y^3$

Now finding the double integral: $\int_0^1 \int_1^2 (x^2 + xy^3)\,dx\,dy = \int_0^1 \left(\frac{7}{3} + \frac{3}{2}y^3\right)dy = \left(\frac{7}{3}y + \frac{3}{8}y^4\right)\Big|_0^1 = \left(\frac{7}{3} + \frac{3}{8}\right) - 0 = \frac{65}{24}$

22.　The volume is given by the integral: $\int_0^1 \int_{x^3}^{x^{1/2}} (4xy - y^3)\,dy\,dx$

Finding the first integral: $\int_{x^3}^{x^{1/2}} (4xy - y^3)\,dy = \left(2xy^2 - \frac{1}{4}y^4\right)\Big|_{x^3}^{x^{1/2}} = \left(2x^2 - \frac{1}{4}x^2\right) - \left(2x^7 - \frac{1}{4}x^{12}\right) = \frac{7}{4}x^2 - 2x^7 + \frac{1}{4}x^{12}$

Now finding the double integral:

$$\int_0^1 \int_{x^3}^{x^{1/2}} (4xy - y^3)\,dy\,dx = \int_0^1 \left(\frac{7}{4}x^2 - 2x^7 + \frac{1}{4}x^{12}\right)dx = \left(\frac{7}{12}x^3 - \frac{1}{4}x^8 + \frac{1}{52}x^{13}\right)\Big|_0^1 = \left(\frac{7}{12} - \frac{1}{4} + \frac{1}{52}\right) - 0 = \frac{55}{156}$$

23. **a.** Finding the average value:

$$\frac{1}{6-0}\int_0^6 28e^{-0.4t}\,dt = \frac{14}{3}\int_0^6 e^{-0.4t}\,dt = -\frac{35}{3}e^{-0.4t}\Big|_0^6 = -\frac{35}{3}\left(e^{-2.4}-1\right) \approx 10.61$$

The average drug concentration is 10.61 mg during the first six hours.

b. Finding the average value:

$$\frac{1}{48-0}\int_0^{48} 28e^{-0.4t}\,dt = \frac{7}{12}\int_0^{48} e^{-0.4t}\,dt = -\frac{35}{24}e^{-0.4t}\Big|_0^{48} = -\frac{35}{24}\left(e^{-19.2}-1\right) \approx 1.458333$$

The average drug concentration is 1.458333 mg during the first 2 days (48 hours).

24. The average value is the double integral: $\dfrac{1}{(10-6)(5-4)}\displaystyle\int_6^{10}\int_4^5 660x^{2/5}y^{3/5}\,dy\,dx = 165\int_6^{10}\int_4^5 x^{2/5}y^{3/5}\,dy\,dx$

Finding the first integral: $\displaystyle\int_4^5 x^{2/5}y^{3/5}\,dy = \frac{5}{8}x^{2/5}y^{8/5}\Big|_4^5 = \frac{5}{8}x^{2/5}\left(5^{8/5}-4^{8/5}\right) = \frac{5\left(5^{1.6}-4^{1.6}\right)}{8}x^{2/5}$

Now finding the double integral:

$$165\int_6^{10}\int_4^5 x^{2/5}y^{3/5}\,dy\,dx = 165\int_6^{10}\frac{5\left(5^{1.6}-4^{1.6}\right)}{8}x^{2/5}\,dx$$

$$= \frac{825}{8}\left(5^{1.6}-4^{1.6}\right)\cdot\frac{5}{7}x^{7/5}\Big|_6^{10}$$

$$= \frac{4,125}{56}\left(5^{1.6}-4^{1.6}\right)\left(10^{1.4}-6^{1.4}\right)$$

$$\approx 3,727$$

The average number of units produced is 3,727 units.